U0245206

生态城市工程规划丛书（第一辑）
Eco-city Project Planning Series (Series 1)

海绵城市与土地构建

Sponge City and Land Construction

［法］安建国　李锦生　钟　律　著
李亚迪　译

序言 1

安建国先生是法国国家注册景观规划设计师，兼具景观能力和艺术专长，博采中法两国文化之长。他1996年毕业于中国鲁迅美术学院；2005年—2009年，他在法国里尔国立高等建筑与景观学院继续深造。这种双重培养训练，培养了他的原创能力，使他成为连接两门学科和两个国家的纽带。

安建国先生著有多篇文章，阐述景观的生态性以及建筑与景观之间的关系。2009年，他与方晓玲共同撰写了第一本图书——《法国景观设计思想与教育》。该书于2012年出版，是第一本系统介绍当代法国景观设计思想与教育状况的著作。他让中国同行了解到：当代中国需要进一步研究和改进景观的研究方式和管理方法，同时，还促进了两国之间的文化交流。

在这本书中，安先生继续他在景观和城市领域的探索与研究——海绵城市。他研究中国和法国城市地域环境建设的不同，并尝试在城市环境中构建雨洪管理和湿地水系，目的是掌握与利用雨水和土地资源，营造和谐的人居环境。

安建国先生积极为法国和中国在景观和风景园林领域的交流搭建桥梁，并做出了卓越贡献。2016年1月，他被列入法国高广艺术、文学与科学院（第十八位院士席）。我们希望这座桥牢不可破，可以促进中法两国之间卓有成效、超越文化的交流，让全世界的专业人士都能受益于这种碰撞出的思想文化。

让–皮埃尔·德戴特
［法］蒙彼利埃大学终身教授
［法］法国高广艺术、文学与科学院院长

Préface 1

Monsieur AN Alexandre Jianguo est un architecte-paysagiste DPLG dont l'originalité tient à la fois à une double compétence dans les domaines de l'architecture et de l'art, et à son acquisition parallèle dans deux pays distincts, la Chine et la France. Diplômé, en art appliqué, de l'Ecole Nationale de Beaux-arts de Luxun, en 1996, il a poursuivi ses études supérieures en France, de 2005 à 2009, à l'Ecole nationale supérieure d'architecture et de paysage de Lille. Cette double formation dans deux pays distincts lui permet de faire le pont, à la fois entre les deux disciplines et les deux pays, une compétence originale et innovante.

Monsieur AN est auteur de plusieurs articles sur le paysage écologique et sur la rencontre entre architecture et paysage. Son premier livre, «Pensées et Enseignement du Paysage en France», rédigé en chinois en 2009, et publié par l'Edition nationale supérieure pédagogique de la Chine en 2012, a été le premier ouvrage contemporain qui présentait le savoir français sur le paysage. Il a permis aux Chinois de comprendre le besoin de développer l'étude du paysage pour une meilleure gestion du paysage et de la qualité de vie, et a favorisé l'établissement d'un échange culturel entre les deux pays.

Dans le présent ouvrage, Monsieur AN poursuit son exploration du domaine du paysage urbain avec le thème particulier de l'Aménagement urbain en éponge. Il envisage une comparaison intéressante des conceptions chinoise et française de l'intégration des milieux humides dans l'environnement urbain, avec pour objectif la maîtrise et l'utilisation des eaux pluviales et de surface, au bénéfice conjoint de l'habitant et du paysage.

Membre correspondant de l'Académie des Hauts Canton depuis Janvier 2016, Monsieur AN a jeté un pont entre la France et la Chine dans le domaine de l'architecture du paysage. Nous souhaitons que ce pont soit le plus solide possible afin de maintenir et promouvoir des échanges culturels fructueux entre les deux pays, et, au-delà, pour faire bénéficier de ce savoir original aux autres professionnels du monde.

Jean-Pierre Dedet
Professeur émérite, Université de Montpellier (France)
Président de l'Académie des Hauts Cantons (France)

序言 2

海绵城市的研究在中国社会的不断认知与探讨过程中获得了新的启发和新的思路，安建国总结了他在法国15年的工作和学习经验，结合中国的具体情况，与业内同仁分享了他对海绵城市的相关研究成果。

作者尽可能使用简单通俗的语言，结合丰富的配图，展示了海绵城市中三个尺度的建设要领。作者坚持海绵哲学，用两种思路尝试寻找对未来城市建设和现状城市改造问题的解决途径。即对于未来城市的建设来说，本着地理区域的宏观—中观—微观的规划设计原则，构建未来良性城市生态环境；对现状城市的建设来说，针对现有城市病特征，结合海绵设计的基础做好老城防涝工作（在暴雨骤降时，除了海绵体继续发挥作用外，也要做好城市范围内紧急疏导内涝的工作，为洪峰流经城市赢得错峰时间）。通过城市“双修”的过程，我国的城市建设走上了良性发展的道路，尊重一切生命，与水为友，可持续性地科学发展。

我们鼓励更多的年轻学者参与祖国的现代化建设，参与学科的良性发展建设，参与国际交流，把爱和职业情怀回报给祖国，唤回青山绿水，唤回灿烂文化，唤回民族自信！

俞孔坚
美国艺术与科学院　院　士
长江学者　特聘教授
国家千人计划　专　家
北京大学建筑与景观设计学院　院　长

Preface Ⅱ

The research on sponge city gains new inspiration and idea through constant cognition and discussion in Chinese society. Alexandre An shares his research results of sponge city by summarizing his 15 years' working and studying experience in France, and combining the specific national conditions of China.

He uses words as simple and popular as possible in combination with various illustrations, to show the main points of sponge city construction in 3 scales. Sticking to the sponge theory, he believes that there are two directions that people can go for the future city construction and the current city renovation, which means that for future city construction, to build optimum urban ecological environment based on the geographic regional macro-middle-micro planning and design principles; for current city construction, to prevent water-logging according to features of current city disease combining sponge design (When rainstorm falls suddenly and heavily, we should urgently dredge water-logging in the city in addition to the continuous effect of sponge, so as to win time for shifting flood peak when it flows through the city). We should lead city construction to a sound development way, respect all the life, make friends with water and develop in a sustainable and scientific way through the "double restoration" process of the city.

We encourage more and more young scholars to join China's construction of modernization, join the sound development and construction of subjects, take part in the international communication, give love and occupational feelings back to the motherland, call back green mountains and clear rivers, glorious culture and national confidence!

Yu Kongjian
Academician of American Academy of Arts and Sciences
Chair Professor of Chang Jiang Scholars Program
Expert of the Recruitment Program of Global Experts
Dean of College of Architecture and Landscape Architecture of PKU

序言 3

伴随着争议和质疑，海绵城市建设自2014年以来一直如火如荼地进行着。如何做好海绵城市设计，一直受到业内外人士的关注和期待。《海绵城市与土地构建》这本书的出版，会像作者期待的那样，使专业人士和公众理解海绵城市建设。

借此机会，笔者和未来的读者交流几个关于海绵城市建设的基本问题，希望能够帮助读者更好地了解本书内容，希望海绵城市建设会得到更多人的理解和支持。

一、海绵城市建设是新生事物吗?

答:“不是!”海绵城市建设是对城市雨水和水系统进行以可持续性管理为目标的设计和建设，这样的工作在世界范围内已经进行了很长时间。在历史上，中国更是积累了非常丰富的水生存经验，这些都是海绵城市建设技术和经验的基础。回顾过去数十年，城市和乡村可持续性水管理更是为世界各国所重视，中国同样产生了一批非常优秀的示范工程项目，很多具有责任感的设计师在自己主持的项目中融进了当今海绵城市建设的工程技术措施。

二、海绵城市建设的性质到底是什么?

答:“改善城市（乡）人居环境。”海绵城市建设工作包罗万象，需要大量资金投入，如果不能产生实际建设效果，势必遭到质疑。海绵城市建设是一个全面提高中国城市人居环境质量的好机会，说明海绵城市建设的这一性质无论对于专业人士还是公众来说都非常重要。城市无论如何建设，终究是为人服务的，只有改善城市（乡）人的居住、出行和游憩环境，公众才会从海绵城市建设工程中直接受益。海绵城市建设工程的综合效益越大，得到公众支持的力度才会越大，这一工作才会持久健康地进行下去。

三、理解海绵城市建设技术措施最大的难点是什么?

答:“不知道水是如何流动的!”为什么这么多城市会突然间面临严

峻的内涝问题，背后无论有多少城市理念、工程方案的失误，都是在就事论事。更加深层次的原因没有那么复杂，只是忽视了例如水从哪里来，水到哪里去，水在城市环境中如何流动，城市建设和管理行为如何影响水的流动，水在流动中会产生哪些变化及效应等问题。这些问题在过去的城市规划、景观设计中都被主观地忽视了，甚至在水工程专业设计中都被简化处理了。这一连串的“疏忽”，还不是直接原因，直接原因是决策者、专业人士和公众都缺乏“水是如何流动的”经验。

四、海绵城市建设做什么？

答：“重建中国城市。”伴随着中国城市发展，似乎在一夜之间，城市内涝席卷南北各地城市。中国本来就是多洪水的国家，在历史上，城市内涝同样频繁，只不过人们对生活品质的追求还没有达到将其上升到“问题”的程度。从未来和今天对城市的要求来看，海绵城市建设是在按照精致化的城市生活与城市形态去改造城市的水系统。从屋顶到街道、道路、广场、绿地，城市的哪个地方没有水呢？海绵城市建设工程毫无疑问是在重新规划中国每个城市，当然，这绝不意味着要把整个城市变成建设工地，而是强调做好海绵城市工程建设必须了解城市乃至整个城市的水系统，进而在适当的地方以适当的规模，通过适宜的技术措施改造城市水系统和人居环境。

完成海绵城市建设这一任务，从来没有像今天这样重要过。本书作者安建国先生以景观设计师（Landscape Architect）的身份撰写此书具有特殊意义。

五、个人如何参与海绵城市建设？

答：“节约用水！”在世界上严重缺水的国家中开展海绵城市建设并不会增加水资源的总量，而节约用水是海绵城市建设获得成功的理念和行动前提，节约用水会是我们对待水所要采取的基本态度与策略。

带着这些基本问题来阅读本书，希望能帮助我们构筑多个尺度、多种愿景的美丽中国蓝图，从书中我们将发现通往未来的技术、设计、思想与行动力量。

李迪华
2017年5月4日于燕园备斋

Preface Ⅲ

Accompanied by controversy and doubt, sponge city construction has been in full swing since 2014. How to do a good job in sponge city design has been focused and expected by people in and out of the industry. The publication of the book *Sponge City and Land Construction* will, like the author's expectation, enable professionals and the general public to understand sponge city.

By this opportunity, I would like to exchange ideas with future readers on a few basic questions on sponge city construction, in order that readers can better understand the content of this book, and that sponge city construction can win more understanding and support.

Question 1: Is Sponge City Construction a New Thing?

Answer: "No!" Sponge city construction is the design and construction of urban rainwater and water systems for the purpose of sustainable management. This work has been in operation for a long time worldwide. China has accumulated rich water survival experience in history, which is the base of sponge city construction technology and experience. In the past few decades, urban and rural sustainable water management has been the focus of countries worldwide. China also produced a number of excellent demonstrative projects. Many responsible designers have been applying today's sponge city construction engineering technical measures to projects designed by them.

Question 2: What Is the Nature of Sponge City Construction?

Answer: "To improve urban (rural) living environment." Sponge city construction includes various contents and requires huge capital investment. If it cannot produce actual construction effect, sponge city construction is bound to be questioned. Sponge city construction is an opportunity to comprehensively improve Chinese urban human living environment quality. Therefore, making the nature of sponge city construction clear is very important to both professionals and the general public. No matter how to be constructed, cities are to serve people. Only to improve living, travel and recreation environment of urban (rural) people, can they directly benefit from sponge city construction projects. The more comprehensive benefits sponge city construction projects produce, the greater support they will get from the general public, so that this work will go on sustainably and healthily.

Question 3: What's the Biggest Difficulty in Understanding Sponge City Construction Technical Measures?

Answer: "Do not know how water flows!" Why so many cities suddenly face severe waterlogging problems? No matter how many mistakes in city concepts and project schemes there are behind these problems, they all scratch the surface. The deeper cause is not complicated: the negligence of such questions as "where the water comes from", "where the water goes", "how the water flows in urban environment", "how urban construction and management behaviors affect the flow of water", "what changes the water produces in flow", and "what effects the water produces in flow", which were subjectively neglected in the past urban planning and landscape design and were even simplified in water project professional designs. This series of "negligence" is not the direct cause. The direct cause is that decision makers, professionals and the general public all lack experience of "how the water flows".

Question 4: What Will Sponge City Construction Do?

Answer: "Rebuild Chinese cities." Along with the development of Chinese cities, it seems that city waterlog swept across northern and southern cities almost overnight. China itself is a country where floods occur. In history, waterlog also occurred frequently in cities, but people's pursuit of life quality did not rate waterlog up to a "problematic" level. Considering the present and future requirements for cities, sponge city construction is to transform urban water system according to the requirement of refined urban life and urban form. From roofs to streets, roads, squares and greenbelts, is there any place without water in cities? Sponge city construction projects are undoubtedly reorganizing each Chinese city. However, this does not mean that a whole city will become a construction site. Instead, it is stressed that good sponge city construction must understand the whole city or even the city water system, so as to transform urban water systems and living environment in appropriate places and scales through appropriate technical measures.

To complete sponge city construction has never been as important as today. The author Alexandre An wrote this book, which has special symbolic significance.

Question 5: How Can Individuals Take Part in Sponge City Construction?

Answer: "Save water!" Carrying out sponge city construction in the world's most water-scarce countries will not increase the total amount of water resource. Water conservation is the ideological and operational premise of the success of sponge city construction. Water conservation will always be the basic attitude and strategy we must take.

With these basic questions, I hope that reading *Sponge City and Land Construction* by Alexandre An can help us to build a blueprint for beautiful China from various scales and visions. In the book, we will find technology, design, thought and action power leading to the future.

Li Dihua

May 4th, 2017 in Beizhai Building, Peking University

前言

首先感谢任南琪院士、俞孔坚、李迪华的启发和指导。

本书的初衷在于探讨和引导海绵城市研究的总体框架和实践方法，梳理海绵城市建设的要领。本人借鉴国内外的成功经验，论述了海绵城市的主要工作内容。希望通过对海绵城市建设思路的探讨，倡导在不同地区有针对性地、深入地研究本地域的系统问题。

书中用简洁易懂的文字和图表阐述海绵城市和土地构建之间的关系，强调学科的综合运用和各工种之间的协作，以及关于地域尺度的海绵城市建设思路；强调在大数据的科学研究后，需要使用者和设计师的场地感知校正，最终完成科学的分析和适当的策划设计；主张人性化和科学化的统一，主张一切生命和谐共处的设计宗旨。

海绵城市建设作为中国新时代环境改善的契机，使我们更准确地理解“Landscape Architecture”的国际发展动态以及寻找适应中国国情的切入点。“Landscape Architecture”在中国的认知和发展经过景观设计或风景园林的阶段之后，最终将以“土地构建设计”的身份承载其在中国发展的历史使命。

海绵城市应该以地域土地研究和地域水系研究为基本依托，因地制宜地设计与实施，通过三种尺度的设计（大尺度的地域雨水径流控制、中尺度的地下管道排蓄疏导和小尺度的城市规划设计），在比城市更大的地域自然环境和水系环境中找到解决方案。设计师只有同时兼顾（包括城市在内的）区域性的土地构建设计和雨洪管理，才能为海绵城市的实现奠定理论和实践基础。

安建国

Foreword

First I would like to express my gratitude to Academician Ren Nanqi, Yu Kongjian and Li Dihua for their guidance.

The original intention of this book is to discuss and guide the general framework and practical method of sponge city research, and sort out the main points in sponge city construction. I have learned from successful experience at home and abroad and discussed the main work of sponge city. I hope to advocate specific and deep study on local system problems in different regions through discussing the ideas of sponge city construction.

In this book, the relation between sponge city and landscape architecture is explained in words and charts which are succinct and easy to understand. It emphasizes the comprehensive application of subjects and the cooperation of different professions, as well as the ideas of sponge city construction in regional scale. It also stresses that after the scientific research on big data, users and designers need site perception to make revision, so as to complete a scientific and analytic research and planning design. It advocates the combination of humanization and science, and the designing purpose of harmonious coexistence of all life.

With the opportunity of environmental improvement in the new era in China, sponge city construction allows us to accurately understand the international development trend of Landscape Architecture and its entry point to adapt to China's national condition. Land Construction Design will finally undertake its historical mission in China's development after its cognition and development going through the stage of landscape design and landscape garden.

Sponge city should be based on regional land research and river system research, and be designed and implemented according to regional condition. It should be solved by the design of 3 scales (large-scale regional rainwater runoff control; middle-scale underground pipeline draining and dredging; small-scale city planning and design) in regional natural environment and river system environment which are larger than cities. Theoretical and practical basis can be laid only through giving consideration to both regional landscape architecture and rain-flood control (including city).

Alexandre An

目录

Contents

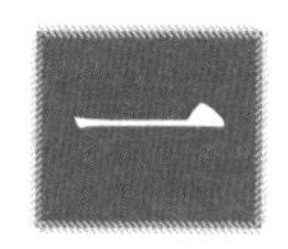

什么是海绵城市?
What Is Sponge City?

海绵城市是指通过加强城市规划建设管理，充分发挥建筑、道路、绿地、水系等生态系统对雨水的吸纳、蓄渗和缓释作用，有效控制雨水径流，实现自然积存、自然渗透、自然净化的城市发展方式。①

“海绵城市”是一个标签，与之相似的另一个标签是“生态城市”。车伍②教授这样解释：简单地打个比方，生态城市如果是一个涉及城市方方面面的多脚生物，那么海绵城市就像是其中的一只脚。

中国工程院院士任南琪对于海绵城市建设和内涵的理解，以及三个建设效益的论述如下：海绵城市应该能够像“海绵”一样，

Sponge city is a development mode that makes full use of ecosystem such as buildings, roads, green lands and water system's functions on absorbing, storing and delaying rainwater by strengthening city planning construction management, so as to effectively control rainwater runoff and achieve natural storage, permeation and purification.①

"Sponge city" is a mark, which is similar to another one named "eco-city". Che Wu② gives a simple example: If eco-city is a multi-foot creature which involves all aspects of a city, the sponge city is just like a foot of it.

Ren Nanqi, an academician of Chinese Academy of Engineering, has his understanding of the sponge city construction and connotation, and 3 benefits of sponge city: The sponge city should be like

---

①摘自《国务院办公厅关于推进海绵城市建设的指导意见》

②车伍（北京建筑大学环境工程系城市雨水系统与水环境教育部重点实验室教授，中国城市科学研究会节能与绿色建筑专业指导委员会委员）

① From *Guidance of the General Office of the State Council on Advancing the Construction of Sponge City*

② Che Wu (a professor of urban rainwater system and water environment, key lab of Ministry of Education of PRC, School of Environment and Engineering, Peking University of Civil Engineering and Architecture; a member of Energy Saving and Green Building Steering Committee of Chinese Society for Urban Studies)

在适应环境变化以及应对自然灾害等方面有良好的“弹性”;重点解决城市涝灾与城市水环境恶化等问题，实现地表水资源、污水资源、生态用水、自然降水和地下水等统筹管理、保护和利用；充分考虑水资源、水环境、水生态、水安全、水文化；减弱热岛效应；确保社会水循环能够与自然水循环相互贯通。

海绵城市建设的实质是城市水资源与水环境综合治理。它会带来社会效益、经济效益和环境效益。

（1）社会效益：通过海绵城市建设和谐宜居、富有活力、各具特色的现代化城市，让人民生活更美好；缓解城市热岛效应，做到小雨不积水、大雨不内涝、水体不黑臭，环境更宜居。

（2）经济效益：减少基础设施建设与维护费用，减轻纳税人负担；提高城市品质、环境承载力和引资能力；开发利用雨水、再生水等非常规水资源，突破缺水地区的发展瓶颈；使海绵城市成为经济发展新的增长点，化解产能过剩的问题。

（3）环境效益：解决城市水体黑臭问题；强化城市污水处理能力；提高城市人均

“sponge”, with good “elasticity” in adapting to environment changes, responding to natural disasters and so on. It should focus on urban flooding, urban water environment degradation and so on, so as to realize the overall management, protection and use of surface water resources, sewage resources, eco-water, natural precipitation and groundwater; fully consider water resources, water environment, water ecology, water safety and water culture; weaken the heat island effect and make sure of the connection between social water cycle and natural water cycle.

The substance of sponge city construction is comprehensive treatment of urban water resources and environment. It will bring social, economic and environmental benefits.

(1) Social benefits: build a harmonious, livable, energetic and featured modern city through sponge city construction, to make people’s life better; ease the city’s heat island effect, and make sure that there is no water logging in light rain and no inner flood in heavy rain; the water is neither black nor smelly, and the environment is livable.

(2) Economic benefits: reduce the cost of infrastructure construction and maintenance to reduce the burden on taxpayers; improve city quality, environment carrying capacity and the ability to attract investment; develop and use unconventional water resources such as rainwater and recycled water to solve the development problems in water-deficient areas; make the sponge city a new growing point of economic development and solve overcapacity.

公园绿地面积和城市建成区的绿地率（图1-1）。

(3) Environmental benefits: solve the black and smelly water problem in the city, strengthen the ability of city sewage treatment and improve the per capita park green area and the green land rate of the build-up area of the city (Fig 1-1).

图 1-1

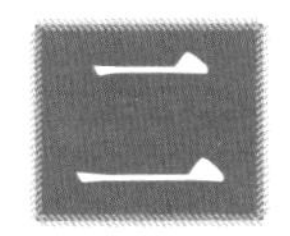

海绵城市的指导原则

Guiding Principles of Sponge City

海绵城市设计的关键不在于城市本身，而在于以土地为设计依托的地域自然环境和水系环境。因此，海绵城市的设计遵循大、中、小三个尺度原则：

·大尺度的地域雨水径流控制（汇水速度、汇水时间、汇水流量）

·中尺度的地下管道排蓄疏导

·小尺度的城市规划设计

海绵城市设计除了要关注城市的雨洪管理之外，还要关注城市外围乡村和郊区的雨洪管理，关注城市之外更大范围的土地问题（图2-1）。李迪华指出，海绵城市设计是一项综合设计，不能单纯地将其认知为城市规划中的防洪排水设计，要站在自然规律的角度上去思考和解决生存方式的问题。海绵城市设计必然是景观设计专业的研究对象，而景观设计专业在海绵城市建设中充当统领的角色。

海绵城市设计涉及的领域非常广泛，

The key of sponge city design is not the city itself, but the regional nature and water environment with land as the basis of design. Therefore, there are 3 principles of scale in sponge city design, which can be divided into large scale, middle scale and small scale.

· Large-scale regional rainwater runoff control (water convergence speed, time and flow)

· Middle-scale draining and dredging of underground pipelines

· Small-scale urban planning and design

In addition to paying attention to the rainwater and flood management of the city, sponge city design should also concern villages and suburbs outside the city, as well as solving land problems out of the city to a larger scale (Fig 2-1). Li Dihua pointed out that sponge city design was a comprehensive design, which should not be just regarded as a flood-control and draining design of city planning. It should think over and solve the problems of living mode from the view of natural rule. Sponge city design must be a subject of landscape architecture, which should lead the practice of sponge city construction.

Sponge city design involves a wide range of fields,

图 2-1

主要有水敏性城市设计（WSUD）、低影响开发（LID）、绿色基础设施（GI）、可持续城市排水系统（SUDS）以及城市修复（UR）等。要做好海绵城市设计，必须对这些领域进行研究和分析。发达国家对于这些领域的研究已经进行了多年，在宝贵的经验中形成了理论知识体系。这些理论是研究海绵城市的重要指导方法。同时，还需要根据本地的自然生态条件，因地制宜地运用和发展这些理论，更好地发挥景观设计师在城市设计和生态设计中的作用，肩负起保护地球资源、协调人与自然关系的责任，在更长远的时空内影响人类对自然的认知与保护。

我国目前出台的相关实践指导主要围绕以LID、WSUD等为代表的西方国家先进的生态雨洪管理技术而展开，虽然愈加关注城市

mainly including water sensitivity urban design (WSUD), low impact development (LID), green infrastructure (GI), sustainable urban drainage system (SUDS) and urban restoration (UR) and so on. To do a good sponge city design, studying and analyzing these fields are necessary. Developed countries have studied these fields for many years, and have valuable experience in theoretical knowledge system, which is important to guide our study of sponge city. Meanwhile, we should use and develop these theories based on local natural ecological condition, so that landscape designers can be the backbone in city design and eco-design, take the responsibility of protecting earth resources and coordinating the relationship between human beings and nature, and influence human beings' understanding and protection of nature in further time and space.

Related practice guidance of our country is mainly based on the eco-rain and flood management technologies of Western countries, represented by

内部排水系统和对雨水的利用、管理，但是在具体技术层面上依旧未能摆脱对现有治水途径中“工程性措施”的依赖。海绵城市的理念不应拘囿于此。它为在不同尺度上综合解决中国城市中突出的水问题及相关环境问题开启了新的旅程，包括雨洪管控、生态防洪、水质净化、地下水补给、棕地修复、生物栖息地营造、公园绿地营造及城市微气候调节等。①

1. 大尺度的地域雨水径流控制

为什么说海绵城市设计的关键不在于城市本身，而在于以土地为设计依托的地域自然环境和水系环境？因为自然降雨并非只在城市范围内，通常是在超出城市范围的地区。这种区域降雨受到诸多自然条件和环境条件的影响，进而对城市产生影响。

在城市之外，我们可以发现不同形态的山川、丘陵、平原、农田、湿地、森林、草地和湖泊；可以发现不同的土壤结构和地质结构；可以发现因地形而产生的微物候环境变化；可以发现城市外围的环境不断受到人为影响的痕迹，如河道裁弯取直、河道两岸硬化、河道缓冲带硬化等。这些客观因素会影响区域降水的河流汇水速度、汇水时间和

①俞孔坚：《海绵的哲学》，载于《景观设计学》第3卷，北京：高等教育出版社，2015年版，第56页。

LID, WSUD, etc., as well as focusing more and more on the draining system and the use and management of rainwater in the city. Technically, it still has not got rid of the dependence on engineering measures of the present flood-control ways, while the idea of sponge city should not be limited to this. It opens a new way to comprehensively solve outstanding water problems and related environment problems of Chinese cities, including water and flood control, ecological flood control, water purification, groundwater recharge, brownfield restoration, biological habitat construction, park green land construction, urban micro-climate regulation and so on.①

1 Large-Scale Regional Rainwater Runoff Control

Why isn't the key of sponge city design the city itself, but the regional nature and water environment with land as the basis of design? This is because natural rainfall happens not only within the city, but always in a larger range out of the city. This kind of regional rainfall is influenced by various natural and environmental conditions, thus having an impact on the city.

Out of the city, we can find different forms of mountains, hills, plains, farmlands, wetlands, forests, grasslands and lakes; find different soil structures and geological structures; find micro-phenological environment changes due to terrain; find the traces influenced by human beings in the environment of city outskirts, such as the curving cut-off of river,

① Yu Kongjian, "*Sponge Philosophy*", *Landscape Architecture Frontiers*, Vol. 3, Beijing: Higher Education Press, 2015, P56

汇水流量，从而对地势较低的城市河道造成汇水洪峰压力。长江两岸的很多城市所面对的主要就是这种区域降雨汇流形成的洪水威胁，而并非城市自身降雨压力导致的洪灾。除非有连续的特大暴雨，才可能出现内涝。

1.1 汇水速度

影响汇水速度的因素主要是高差和水体流经的地表形态。陡峻的山坡汇水速度相对较快，平缓的山岭汇水速度相对较慢。

水体流经光滑的地表（例如，水泥硬化的河道）时，水流速度增大，同时水体的冲击力和破坏力增大（图2-2）；水体流经阻挡物多的地表（例如，石头和植被较多的河道）时，速度会变慢，水体流经地表弯曲河道的速度明显比流经笔直河道的速度慢（图2-3）。

四川青城山后山泰安古镇的震后恢复建设中，穿越古镇的味江河河道设计就出现了

river banks and river buffer zone hardening, etc. These objective factors will influence river water convergence speed, time and flow of regional rainfall, and thereby bring about flood peak pressure on city river in lower places. The main flood threat of many cities along both sides of the Yangtze River comes from this kind of rain water convergence, but not from the rainfall of the city itself. Water logging may only happen due to continuous heavy rain.

1.1 Water Convergence Speed

Water convergence speed is mainly influenced by height difference and the surface morphology that water flows through. Water converges very fast in steep hillsides and relatively slowly in gentle slopes.
Water flows faster when flowing through smooth surface(such as river courses with hardened cement), with the impact and destructive power of water being stronger (Fig 2-2). Water flows slower when flowing through surface with many obstructions(such as stones and plants near the river). When water flows through curved river courses, the speed is obviously lower than that of straight river courses (Fig 2-3).

图 2-2

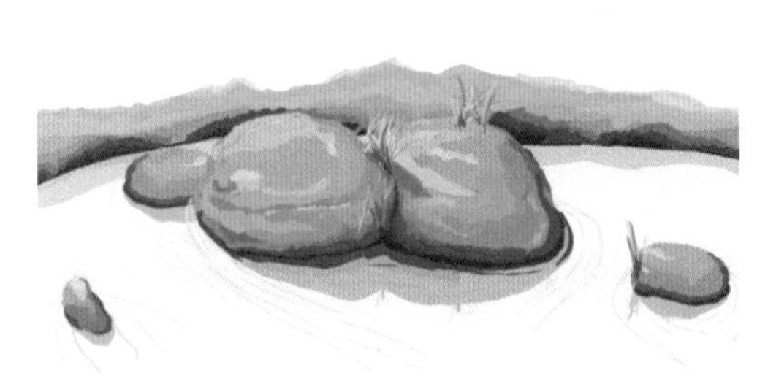

图 2-3

问题。

1900年，河道状态自然，人口稀少，耕地围绕着河道。

1950年，河道仍然保持自然状态，人口有所增加，新增人口的住房主要集中在味江河的“儿”字形缓冲带（图2-4）。

2000年时的河道变化很大，河道硬化变窄，人口急剧增加，且新增人口住房占用河道，河道缓冲带全部硬化，河道中心建了一座人工岛，耕地大量减少。此时，味江河的生态环境已经开始恶化，从未出现过的洪水问题开始频发。

2008年汶川地震时，泰安古镇的人口已经相当密集，而耕地面积持续减小。震后的恢复建设没有改变河道硬化的局面，反而增大了河道硬化的面积，甚至硬化了河床，导致河水无法与河床土壤进行氧气和有机物的交换，生物栖息地被严重破坏。钢筋水泥在

In the after-earthquake reconstruction of Tai'an ancient town, back of Qingcheng Mountain, the river course design of Weijiang River, which flows through the old town, has problems.

In 1900, the condition of the river course was natural. The place was sparsely populated, and farmland surrounded the river course.

In 1950, the condition of the river course was still natural. The population increased, and the house construction of added population was mainly concentrated in the buffer zone of Weijiang River, which looked like the Chinese character "儿" (Fig 2-4).

In 2000, the river course changed a lot. It was hardened and narrowed. The population increased dramatically. What's more, the houses occupied the river course. The buffer zone of the river was totally hardened. An artificial island was built in the center of the river, and there was a substantial reduction in farmland. Meanwhile, the ecosystem of Weijiang River had started to deteriorate with frequent flood, which never happened before.

In 2008, when earthquake happened in Wenchuan,

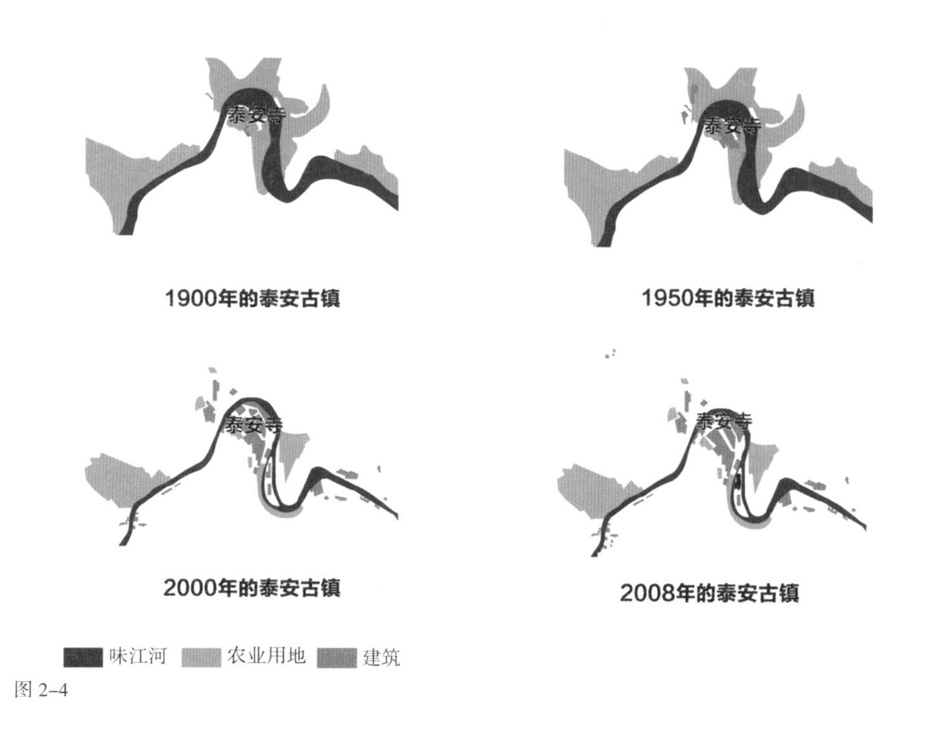

图 2-4

奔流不息的河水面前是多么脆弱！仅仅两年的时间，河底硬化的水泥已经被河水摧毁。硬化的河岸切断了人们与河水的联系，这条河流已经失去了它的价值。此外，以前河道之中错落着大大小小的卵石，总有孩子和游客在河中的石块上休息，聆听水声。如今，这些卵石在河道治理中全都被清除，取而代之的是沉积下来的淤泥，曾经清澈的河水不复存在。河水为何清澈？就是因为水与卵石撞击，翻起浪花，带走了淤泥，留下美丽的卵石和清澈的河水（图2-5）。

类似情况在全国各地屡见不鲜，不仅破

the population of Tai'an ancient town was quite dense, while the farmland continued to decrease. The restoration of construction after the earthquake did not change the hardened river course; on the contrary, the riverbed was hardened. As a result, the river water could not exchange oxygen and organic matter with the riverbed soil, and the biological habitat was badly damaged. How fragile the reinforced concrete was when facing the flowing river water! After only 2 years, the hardened cement under the river was destroyed by the river water. The hardened river bank cut off the connection between people and river water, and the river had lost its value. Moreover, in the past pebbles of different sizes scattered in the river; children and tourists always rested on stones in

图 2-5

坏了生态环境，也增加了洪水和汇水过快造成的威胁，给下游城市带来洪灾压力。

图2-6是波尔多公园的步行道设计，坡度约为8°，石材在草坪上呈线形铺装。草坪可以直接吸收步行道路汇集的雨水，线形铺装的石材也可以拦截流动的雨水，使雨水缓慢下渗，防止水土流失。

通过设计减缓城市外围更大尺度流域的汇水速度对海绵城市有很多益处：

（1）减少水土流失，保护地表生态环境和生物多样性；

（2）错峰，延迟洪峰到达城市的时间，为排除城市内涝赢得时间。因为洪峰到来时，若城市内涝无法排除，则只有等到洪峰退去，方可排涝；

（3）降低水体快速通过时所带来的破坏。

the river; listening to the sound of the flowing water. But now, these pebbles have been cleared away and replaced by deposited silt. The clear river has gone forever. Why was the river clear? Because water hit pebbles and waves rolled, taking silt away and leaving beautiful pebbles and a clear river behind (Fig 2-5).

This kind of failure is common nationwide, which not only damages ecosystem, but also increases the threat of flood and water convergence speed, which may cause flood in downstream cities.

Fig 2-6 shows the trail design in Bordeaux Park, with a slope of about 8° and stone materials being linearly paved on the lawn. The lawn can directly absorb rainwater collected from the trail and the linearly paved stone materials can also stop flowing rainwater, so as to make rainwater infiltrate slowly and prevent water loss and soil erosion.

Slowing down the water convergence speed in a larger scale of city periphery by designing brings a lot of benefits to the sponge city:

(1) Reduce water loss and soil erosion, and protect surface ecological environment and biological

图 2-6

1.2 汇水时间

影响汇水时间的因素主要是土壤结构。好的设计能让雨水下渗，通过地下渗水的方式汇入河流。这样，一部分雨水通过渗流的方式延长了汇水时间，也延迟了洪峰到达城市的时间，同样为排除城市内涝赢得了时间。

汇水时间的控制和设计是海绵城市项目无法在城市自身范围内解决的问题。必须着眼于比城市范围更大的地域尺度水系和国土尺度水系，才能顺利地控制城市水系。

1.3 汇水流量

影响汇水流量的因素主要是降雨量、汇水速度和汇水时间。特定时间、特定河道区

diversity;

(2) Avoid flood peak, put off the time that the flood peak reaches the city, and meanwhile win time for the city to get rid of water logging, since it cannot be removed when the flood peak reaches and draining only can be done after the peak;

(3) Reduce the damage caused by water when it passes quickly.

1.2 Water Convergence Time

Water convergence time is mainly influenced by soil structure. Good design can make rainwater permeate into the ground thus flowing into river. Thus, the convergence time of part of rainwater is prolonged by permeating, and the time of flood peak reaching the city is delayed. This also wins time for draining water logging of the city.

The sponge city project cannot solve water

段的汇水流量决定了洪灾的危害程度。因此，设计和管理汇水速度和汇水时间对控制汇水流量、减小洪水危害具有重要意义（图2-7）。

对大尺度地域雨水和水系的控制直接影响海绵城市的治理效果。这项工作要联合环境保护部门、水利部门、林业部门和农业部门等协同推进，形成一个以地域及国家综合发展为目标的良性思考和运营机制。

2. 中尺度的地下管道排蓄疏导

城市内涝的产生有两个主要原因：

一是暴雨骤降，几小时之内甚至会达到全年降雨量。这种情况下，瞬间出现的城市内涝必须以最快的速度排出，除了依靠地表径流排出之外，城市管道排洪也起到不可替代的作用。

convergence time control and design within the city itself. To control urban water system more smoothly, designers must focus on regional and national water systems, which have a larger scale than urban water system.

1.3 Water Convergence Flow

Water convergence flow is mainly influenced by rainfall, water convergence speed and time. Water convergence flow of a specific river section in a specific time decides the degree of flood damage. Therefore, designing and managing water convergence speed and time are significant to control water convergence flow and flood damage (Fig 2-7).

Controlling large-scale regional rainwater and water system directly affects the effect of sponge city management. This should be carried forward cooperatively by the departments of environment protection, water conservancy, forestry and agriculture, etc. to form a good thinking and operating system of regional and national comprehensive development.

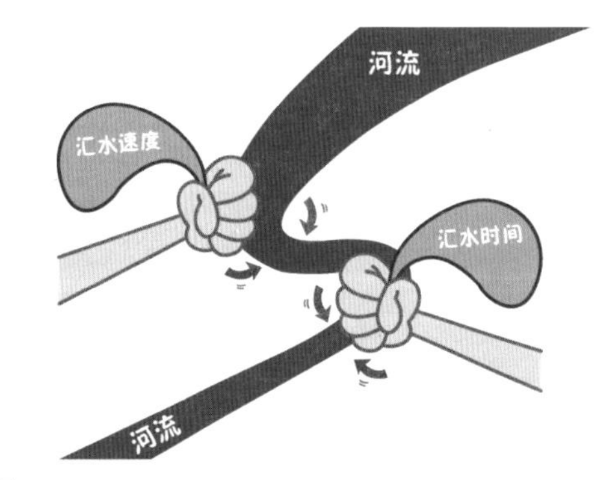

图 2-7

二是城市原有排水管网过细造成淤堵，无法应对暴雨骤降时的内涝排放，最后因为排放时间过长而错过了最佳的内涝排放时间。当上游洪峰到来时，河道水位高，河水倒灌，无法排涝，在城市内形成无法根治的水患。只有等到洪水过后，河道水位下降，内涝才能排出（图2-8）。

因此，中尺度的海绵城市设计重点在“应急排水”上。在应急排水的同时，城市内的调蓄池和海绵体继续发挥调节水量、降低洪水威胁的作用。在暴雨连续骤降的情况下，不要等到城市调蓄池和海绵体饱和后再进行排水，那时内涝和水灾将无法控制。

水利部门应该根据各个地区不同的降雨量和土壤地质环境，因地制宜地研究制定适于本地区的排洪管网标准，避免造成排水功能上的失误以及资金浪费。高标准的城市管道设计能有效应对暴雨骤降时的内涝排放问

2 Middle-Scale Draining and Dredging of Underground Pipeline

There are 2 main reasons for water logging within the city.

The first is the rainstorm, with the rainfall even reaching the amount of the whole year in several hours. In this situation, city water logging that instantly appears must be drained with the fastest speed. To drain the water, the role of city pipelines is irreplaceable besides surface runoff.

The second is that the original drainage pipelines are too thin and have been blocked. As a result, they cannot drain water when rainstorm comes. At last, the best drainage time for water logging is missed due to the long drainage time required. When the upstream flood peak comes, the water level of the river becomes high and the river flows backward, so drainage becomes impossible and flood in the city becomes uncontrollable. The water logging can be drained only after the flood is over and the water level drops (Fig 2-8).

Therefore, the key point of middle-scale sponge city design is “emergency drainage”. While draining

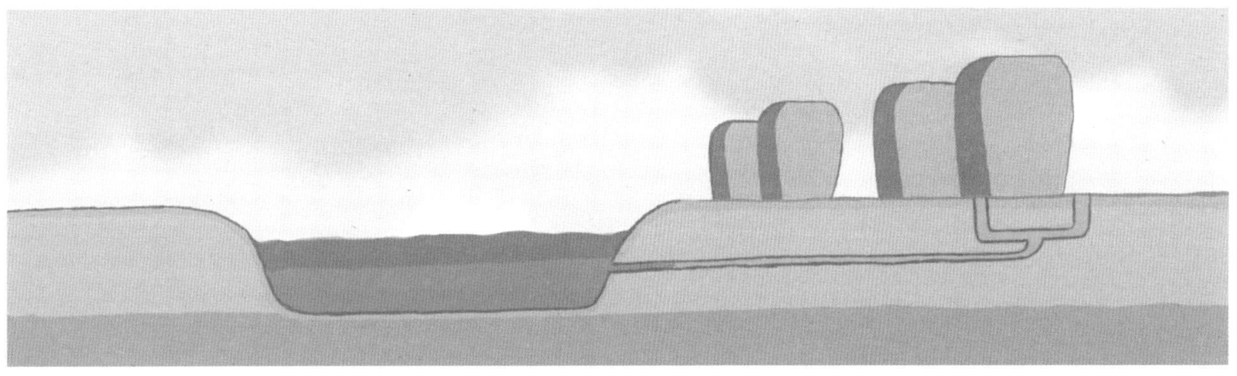

图 2-8

题（图2–9）。

3. 小尺度的城市规划设计

《国务院办公厅关于推进海绵城市建设的指导意见》（国办发〔2015〕75号）指出：通过海绵城市建设，综合采取“渗—滞—蓄—净—用—排”等措施，最大限度地减少城市开发建设对生态环境的影响，将70%的降雨就地消纳和利用。到2020年，城市建成区20%以上的面积达到目标要求；到2030年，城市建成区80%以上的面积达到目标要求（图2–10）。

为什么要对海绵城市建设采取上述六种处理手法?

（1）城市绿地面积小、硬化面积大，雨水无法下渗，致使汇水流量增加，增大了河道的洪峰压力；

emergently, the regulating pool and sponge in the city still regulate the amount of water and reduce the threat of flood. When continuous rainstorms come, we cannot drain water after the regulating pool and sponge are full, otherwise water logging and flooding will be inevitable.

Water conservancy department should research and formulate the standard of flood draining pipeline according to the local rainfall and soil geology environment, so as to be suitable for the local region and avoid mistakes of drainage function and waste of money. High-standard urban pipeline design is effective on water logging drainage when there is rainstorm (Fig 2-9).

3 Small-Scale City Planning and Design

In the *Guidance on Advancing the Construction of Sponge City* (No.〔2015〕75 *document issued by the General Office of the State Council*): Comprehensively taking measures of “permeating—delaying—storing—purifying—using—draining”

图 2–9

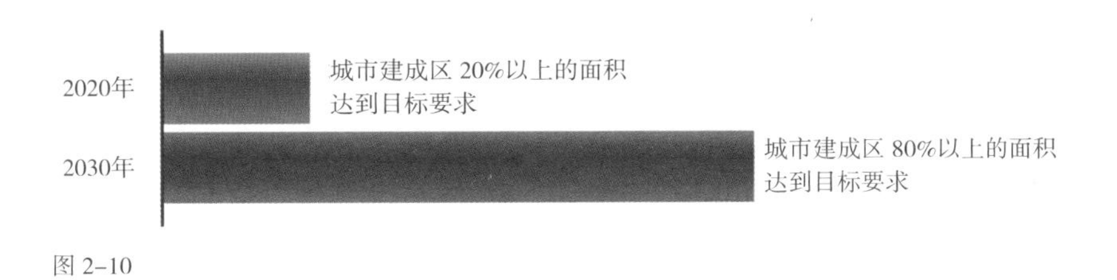

图 2-10

（2）城市中大片高层建筑的地基切断了树木根系之间的联系，导致地下含水层下降（图2-11）（树木根系之间的联系形成网络，承载着地下含水层，并且使地下含水层接近地表，调节地表湿度和温度，提供必要的水源）；

through the construction of sponge city to reduce the impact of urban development and construction on ecosystem to the full extent, as well as store and use 70% of rainfall on the spot. Till 2020, over 20% of the built-up urban area should meet the requirements and over 80% in 2030 (Fig 2-10).

Why should we take the above 6 measures on sponge city construction?

图 2-11

（3）部分地表绿地无法成为真正有效的海绵体（没有下渗途径），如城市地下停车场上方的绿地（图2–12）。

(1) As a result of small green area and big hardened area in the city, rainwater cannot permeate and the amount of confluence increases, which adds more pressure of flood peak on the river;
(2) In the city, the foundations of many high-rise buildings cut off contact among tree roots and lead to the decline of underground aquifer (Fig 2-11). (The contact among tree roots forms nets, which carry the underground aquifer and make it closer to the land surface to adjust the humidity and temperature of land surface and provide necessary water source);
(3) Part of the green surface cannot actually become effective sponge (since there is no way of permeating), such as the green land above the underground parking lot (Fig 2-12).

图 2–12

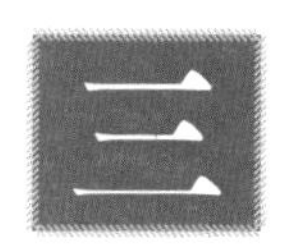

海绵城市设计与土地构建设计的关系

The Relationship Between Sponge City Design and Landscape Architecture

海绵城市解决的是生存环境问题，海绵城市的设计和实践旨在改善城市居民的生存环境。海绵城市设计与土地构建设计密切相关。它包括两种关系的处理，即人与自然环境的关系和人与生存之地的关系。这是欧美学者对景观设计研究最敏感的主题之一。早在19世纪，美国人奥姆斯泰德创造了“Landscape Architecture”一词，其中Land是土地，Scape是景色，Architecture是构建（在此并非建筑之意），因此，Landscape Architecture 指土地上一切元素的构建形成了景色。

该词的重点含义有三个：

（1）土地；

（2）一切元素（包括土地上的一切生命）；

（3）构建关系。

19世纪该词的创造的确有前瞻色彩。但是，欧洲人在当时对这一词的理解是非常含糊的，甚至直到20世纪70年代，欧洲学者对

The current sponge city is to solve the problem of living environment, aiming at improving people's living environment through the design and practice of sponge city. Sponge city design is closely related to Landscape Architecture design. It deals with two kinds of relationships, that is, the relationship between human beings and nature and the relationship between human beings and the land for living. This is one of the most sensitive themes of landscape architecture that European and American scholars are concerned with. As early as the 19th century, an American Olmsted created the word “Landscape Architecture”, concerning Land, Scape and Architecture, which means that the scape is architected by all the elements on the land.

This word has 3 main meanings:

(1) Land;

(2) All the elements(including all the lives on land);

(3) Architecting relationships.

When invented in the 19th century, it was really forward-looking, but Europeans' understanding of this word was very ambiguous. Even in the 1970s, European scholars still regarded Landscape

Landscape Architecture的理解仍是“糊里糊涂一锅粥”。

这种评论源自何处？事实上，在Landscape Architecture一词被提出后并没有形成因社会需求所引发的推动力，也没能推动该学科的完善和发展。

在美国，先知先觉的奥姆斯泰德成了行业探索者，这也正反映了美国英雄主义的民族特征。在欧洲，人们一直在等待Landscape Architecture出现的理由，因为以法、德为主导的欧洲价值观与美国截然不同。欧洲人更关注的是生存状态的反馈和社会群体的需求，他们更关注平等对话和交流的感觉。因此，Landscape Architecture在欧洲的真正诞生是欧洲社会发展和需求推动的结果。

在欧洲，Landscape Architecture是如何出现的？

欧洲地理学家对土地和人居环境的研究为Landscape Architecture在欧洲的诞生奠定了近代理论基础。

19世纪地理科学的发展是西方现代精神最显著的特点之一，尤其是在法国和德国。Landscape Architecture的第一层含义涉及大地表面，因此，地理科学是解释景观物理特征的第一门学科。

欧洲近3000年的地理科学研究经历了从“实证地理学”到“人文地理学”的进步过

Architecture as “a mess”.

Why was there such a kind of comment? In fact, it was because after the word Landscape Architecture came out, there was no driving force resulting from social needs to promote the completion and development of it.

In America, Olmsted became the explorer of the industry, which just reflected the American ethnic characteristics of heroism; while in Europe, people were waiting for the reason of its coming, because the value of Europe led by France and Germany was totally different from that of America. Europeans paid more attention to the feedback of survival state and needs of social groups. They paid more attention to the feeling of equal dialogue and communication. Therefore, the real emergence of Landscape Architecture in Europe was the result of the promotion of social development and needs.

In Europe, how did Landscape Architecture appear?

European geographers’ study on land and living environment laid the modern theoretical foundation for the emergence of Landscape Architecture in Europe.

The development of geographical science in the 19th century was one of the most outstanding features of Western modern spirit, especially in France and Germany. The first meaning of Landscape Architecture involved land surface; therefore, geographical science was the first subject that explained physical characteristics of landscape.

Over the 3,000 years, European geographical science study has experienced the improving process from

程，该进步的转折点就在19世纪。

19世纪初，地理学逐渐发展成为能够清晰解释人与其生存环境关系的一门学科。这里的生存环境指自然环境，但是一旦涉及人与其生存之地的关系，研究就陷入僵局，这就是实证主义地理学的特征。这种实证主义地理学与社会科学格格不入，促使一些地理学家转向文化主义或者地区地理学来寻求出路。

代表人物是法国著名地理学家维达尔·白兰士，他认为："人的行为对一个地方有特殊影响。"受到其德国老师洪堡、拉采尔，特别是李特尔的影响，白兰士更重视居民与其生活基地之间关系的研究，他创造的人类地理学虽然仍然属于自然科学的范畴，但避免了僵化的达尔文狭义环境主义。事实上，19世纪地理科学由实证地理学发展到人文地理学的过程成为景观设计学在欧洲发展的坚实理论基础。

20世纪50年代，华盛顿大学的教授爱德华·L·乌尔姆认为："学院派地理学的弱点在于它忽略了两个最先锋研究者触及的问题，特别是白兰士的研究工作。人与其生存地之间的关系虽然得到了重视，但是沟通问题却被忽略。"爱德华因此提醒年轻的研究工作者重视存在的"关系网络"。[①]

① 安建国，方晓灵：《法国景观设计思想与教育》，北京：高等教育出版社，2012年版，第15页。

"positivist geography" to "human geography", and the turning point of the improvement was just in the 19th century.

At the beginning of the 19th century, geography developed gradually as a subject that could clearly explain the relationship between human beings and their living environment. Here, "living environment" referred to natural environment, but when it came to the relationship between human beings and their living land, the study went to a dead end. This was the feature of positivist geography, which was out of tune with social science, and promoted some geographers to turn to culturalism or regional geography to seek a way out.

The representative figure was Vidal de la Blache, a famous French geographer. He thought that human behavior had special effects on a place. Influenced by his German teachers Humboldt, Ratzel and especially Ritter, Blache attached importance to the study of relationship between residents and their living base. He created human geography, which avoided the inflexible Darwin's Narrow Environmentalism, although it still belonged to the scope of the natural sciences. In fact, the process of geography in the 19th century developing from positivist geography to human geography became the solid theoretical basis of the development of Landscape Architecture in Europe.

In the 1950s, professor Edward L Ulm of the University of Washington thought that the weak point of academic geography was that it ignored

直到20世纪60年代，欧洲新一代的地理学家提出将实证地理学和人文地理学融为一体的可能。于是，一个重要的时刻来临了，1957年在法国南锡召开的“欧洲乡村景观研究系列研讨会”上，历年来的地理发现和假设汇集产生了碰撞。之后，安德·美涅在1958年发表了总结性著作《农耕景观》（或译成《生产性景观》）。虽然景观没有被明确定义，但是关于景观研究工作的地位已被确认。

总体来说，这个时期的法国地理学家主要有两个研究方向：（1）空间经济学的发现使大多数法国地理学家朝着经济学家想象的理论模式方向深入；（2）另一些地理学家则发展人文的观点。因此，1976年乔治·贝纳德建议在结合自然（地理系统）、社会（地域）和文化（人文）三个研究方向的基础上做一次总结。这种综合学科的发展思路奠定了近代景观设计学教育思想和实践体系的基本框架。

随后有诸多学者在地理学、社会学、人文学、心理学、法学、建筑学、艺术学、规划学等多个领域展开对景观的深入研究。直到20世纪末，人们就生态环境改造的观点达成共识并取得明显成效之后，景观才被推到精神层面和感知层面的研究上，设计师才意识到不可见景观和可见景观都是设计的重要

the problems that two pioneer researchers studied, especially Blache's work. Although the relationship between human beings and living land was emphasized, communication was ignored. Therefore, Edward reminded young researchers of paying attention to "relationship network".①

Until the 1960s, the new generation of European geographers put forward whether positivist geography and human geography could be integrated. Then an important moment came. In 1957, European Rural Landscape Research Seminar Series was held in Nancy, France, and geographical discoveries and assumptions over the years were put together and discussed. After that, André Meynier published a summary book *Farming Landscape* in 1958. Landscape was not clearly defined, but the status of landscape study was confirmed.

Generally speaking, in this period, French geographers had 2 research directions. First, with the discovery of space economics, most French geographers furthered their study toward the theoretical model that economists imagined. Second, others developed humanistic points of view. Therefore, in 1976, George Benard suggested making a summary based on the combination of nature(geographical system), society(region) and culture(humanities). This development idea of comprehensive subject built the basic framework of

① An Jianguo and Fang Xiaoling: *Thought and Education of Landscape Architecture in France*, Beijing: Higher Education Press, 2012, P15.

内容。

地理科学推动了景观设计理论研究的发展，但欧洲社会发展的内在因素最终决定了景观设计学科的形成。例如，德国景观设计学科的形成受到工业革命的影响。20世纪60年代，德国宣布工业革命结束，昔日最为辉煌的西部工业区瞬间冷清下来，失业人口暴增。德国政府面临着巨大的经济压力和社会转型压力。德国西部的主要工业城市必须通过转型来适应这一社会巨变，即工业城市向金融城市转型、工业城市向旅游城市转型、工业城市向文化城市转型。在这个转型过程当中，没有一个行业可以承担转型的工作重任，德国需要推出一个新的综合专业来应对这一时代巨变，而当时的景观设计适应了这一巨变并扮演着重要的角色。

例如，自1945年第二次世界大战结束以后，法国的城市化进程持续了20多年。城市化进程速度之快，迫使法国必须从原殖民地国家吸纳大量的外来劳动力。于是，近300万的劳动力涌入法国，聚集在大城市周边，主要从事城市基础设施建设和房地产开发工作。

当时在法国大城市郊外出现了很多豆腐块儿似的劣质居民楼，直到今天法国还在对其陆续地拆迁更新。20世纪60年代末，法国经历了轰轰烈烈的城市化运动之后，巴黎人开始思考巴黎的身份特征。人们从机场

education thought and practice system of the modern landscape architecture.

After that, many scholars furthered their study on landscape in the fields of geography, sociology, humanities, psychology, law, architecture, art and planning, etc. Until the end of the 20th century, after the point of ecological environment transformation was widely recognized and achieved remarkable results, landscape study was pushed to the level of spirit and perception. Designers felt that both invisible and visible landscapes were important design content.

Geographical science promoted the development of study on landscape architecture theory, but it was internal factors of European social development that finally promoted the formation of landscape architecture. For example, the formation of landscape architecture in Germany was influenced by the industrial revolution. In the 1960s, Germany announced that the industrial revolution was over and the most splendid West Germany industrial zone was deserted in a moment with the number of unemployed population increasing sharply. German Government faced huge economic pressure and social transformation pressure. Main industrial cities in West Germany had to adapt to this social change by transformation, including the transformation from industrial cities to financial cities, tourism cities and cultural cities. In this process, there was not an industry that was able to take the heavy responsibility of transformation, so Germany needed to put forward a new and comprehensive subject to respond to this

到市中心，看到的是琳琅满目的广告牌，但这些广告牌不是巴黎的身份特征，也不是巴黎的城市形象。那么，巴黎应该是什么样的？与此同时，城市与乡村交界的部分成为无人关注的灰空间，城市间的交通网络设计也被忽略。面对这些特殊的时代问题，不论是建筑师、规划师、艺术家，还是园林设计师都无法从整体和宏观角度对这些问题进行梳理和解决。正是在这种社会需求的背景下，法国才有了景观设计专业。

Landscape Architecture在中国是如何出现的？

在中国，学术界将其译为“景观设计”或“风景园林”。笔者认为在未来的发展中，它应该被译作“土地构建设计”。

20世纪90年代初，景观设计开始成为政府形象工程和“城市美化运动”的代名词。与此同时，三类（农林类、建筑规划类和艺术类）院校也纷纷开设了景观设计专业。各类学院基本都在本院擅长的知识体系和角度下诠释Landscape Architecture。

伴随着中国城市化运动，景观设计经历了近20年的火爆招生和就业之后，如今似乎进入了行业瓶颈期。但回首中国景观设计学的发展历程，我们并没有建立起属于自己的景观教育体系，大多数院校还是处于各自为

great change of the times. Landscape architecture was adapting to this transformation change and played an important role.

For example, in 1945, after the Second World War was over, the urbanization process in France lasted for more than 20 years. The fast speed of urbanization forced France to absorb a large number of foreign labors from original colonial countries. Hence, about 3 million labors swarmed into France, gathered around big cities and mainly engaged in infrastructure construction and real estate development.

At that time, many low-quality residential buildings appeared on the outskirts of big cities of France, which are still being removed and rebuilt today. At the end of the 1960s, after intense urbanization movement, Parisians started to consider the identity characteristics of Paris. People saw dazzling billboards from the airport to the city center of Paris, but these billboards were neither the identity characteristics nor the image of Paris. Then what should Paris look like? Also, the urban and rural border became unattended gray space, and the traffic network design among cities was ignored. Facing these problems in special times, architects, planners, artists and garden designers could not sort and solve them comprehensively. Just in this background of social needs, landscape architecture came into being in France.

How did landscape architecture appear in China?

In China, scholars translated it into “landscape design” or “landscape garden”. The author of the book thinks that in future development, it should be

政的探索式教学阶段。而这一时期特殊的政府形象工程和“城市美化运动”也不适合作为中国景观设计理论研究的基础依据。因此，中国景观设计的教育体系研究和实践方法研究还有很长的路要走。

真正的中国景观设计应该是在“海绵城市”时代到来时诞生的。因为这个时代具备了中国景观设计诞生的理由，这个理由很简单，就是老百姓要喝一口纯净的水、吸一口清新的空气，在宜居的环境中居住。这些社会需求为这个学科的发展提供了良好的土壤。

中国在“海绵城市”时代，主要工作还处于“人与自然环境的关系”的研究阶段。在未来，当中国进入“人与生存之地的关系”的研究阶段时，就进入了土地构建设计阶段，那时“景观设计”就自然而然地变成了“土地构建设计”。

在2012年出版的《法国景观设计思想与教育》一书中，作者对景观设计（Landcsape Architecture）给出了新的定义，即“景观设计”是以土地为依托、以时间为脉络、以自然的自我管理为特征、以使用者的体验为论证依据的现代科学，它研究一切生命的和谐共存关系。在未来将以“土地构建设计”这一名词来解释其定义。

图3-1显示了景观设计（土地构建设

translated into “landscape architecture”.

From the beginning of the 1990s, landscape architecture became the synonym of both the government image project and “city makeup movement”. At the same time, 3 categories of institutions (agriculture and forestry, architecture planning and art) set up landscape architecture major in succession. They generally explained landscape architecture within the knowledge system and viewpoint that they were good at.

With the growth of urbanization movement in China, landscape architecture seems to enter the period of stagnation today after nearly 20 years of flourishing enrollment and employment. Looking back to the process of landscape architecture, we did not really build our own landscape education system and most institutions were still in the period of exploratory teaching. The special government image project and “city makeup movement” in this period were not suitable to be the basis of study on landscape architecture theory in China. Thus, the study on education system and practical methods of landscape architecture in China still has a long way to go.

The real birth of China’s landscape architecture should be from the arrival of “sponge city” era, because this era has the reason for its birth. Such kind of reason is very easy. People wish to drink clean water, breathe fresh air and live in a livable environment. These social needs provide a good basis for the development of this subject.

In current “sponge city” era, China’s main work is still in the period of studying “the relationship

计）与城市规划设计和乡村规划设计的关系。

为什么景观设计要跨越到土地层面，最终进行土地构建设计？

因为土地是承载万物的基体，如今的景观设计已经跨越了空间和时间的界限，在不同尺度下进行生存方式和生态环境的研究。

例如，巴黎贝西新区的“昆虫旅馆”设计（图3–2）。这项设计由若干木块（用电钻钻出小孔）拼凑而成，这些孔洞就是为了使像小蜜蜂这样的传粉昆虫有躲藏的地方。

为什么要设计可以供传粉昆虫躲藏的地方呢？因为在巴黎、北京和上海这样的大城市，公园绿地之间的距离通常大于 5 km，而传粉昆虫的飞行距离只有 2.5 km~3 km。小蜜蜂飞行 2.5 km 后会疲劳，它们需要休息，那

between human beings and natural environment". In the future, when China begins studying "the relationship between human beings and the land for living", it will enter the period of Landscape Architecture, then "landscape design" would become "landscape architecture".

In the book *Thought and Education of Landscape Architecture in France* published in 2012, the author of the book gave a new definition: "landscape architecture" is a modern science that is based on land and characterized by self-management of nature, taking time as context and user experience as an evidence basis. It studies the harmonious coexistence of all lives. In the future, the term "landscape architecture" will be used to explain its definition.

Fig 3-1 shows the relationship between landscape architecture and urban and rural planning.

Why should landscape architecture go across to the level of land, and finally become land design?

That's because land is the basis that carries

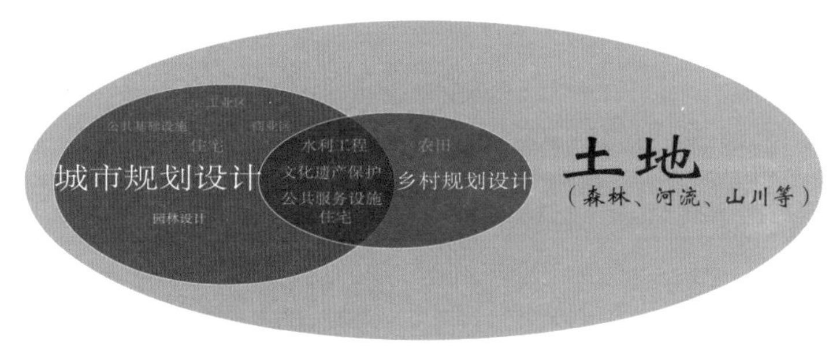

图 3–1

图 3–2

么在休息的过程中，如果不受到保护很可能被上一级动物捕食。设计师设计“昆虫旅馆”就是为了让蜜蜂这样的传粉昆虫在进行传粉采蜜的时候有一处安全的栖身之所（图 3–3）。

这项设计看起来虽小，但搭建起了大城市绿地和公园之间的生态系统构架。这个案例说明景观设计的工作重点不在于一个绿地植被的塑造，而在于整个生态系统的协调构建。而这些基础的生态系统构建有利于人类安全地介入生态系统，参与生态系统的良性运转。

这项设计虽然是为小昆虫而做，但其背景是以土地为依托的地域生态系统设计。

图3–4是在法国随处可见的城市绿地，一部分草坪被修剪，还有一部分未被修剪。

那片未被修剪的草坪有何用意？这是为

everything. Today, landscape architecture has stepped over the limit of space and time, and becomes a research on the way of life as well as ecosystem in different scales.

For example, the design of “Insect Hotel” in Bercy(Fig 3-2), Paris. It is made of several wood pieces with electric drill drilling holes on them. These holes are for pollination insects such as bees to hide in.

Why should they design places for pollination insects to hide in? That’s because in metropolises such as Paris, Beijing and Shanghai, distance between park green lands is always longer than 5 km, but the physical flight distance of pollination insects is only 2.5 km to 3 km. Bees become tired and have to rest after flying 2.5 km. While resting, if these bees cannot be protected, they are very likely to be eaten by higher-level animals. Designers designing “Insect Hotel” is to give pollination insects such as bees a

图 3-3

safe shelter when they are pollinating and gathering honey (Fig 3-3).

This design seems to be small, but it completes an ecosystem architecture between green land and park in big cities. This case illustrates that the main point of landscape architecture is not shaping green land vegetation, but the coordinated construction of the whole ecosystem. Furthermore, these basic ecosystem constructions are good for human beings to safely get involved in ecosystem and take part in the good operation of ecosystem.

Though this design is for small insects, its background is regional ecosystem design based on land.

图 3-4

了给城市的昆虫提供一处栖身之所。

在今天的大城市中，人们很少能够看到蝴蝶了，它们都到哪里去了呢？蝴蝶主要在植物叶面上产卵，当城市的园丁把草坪剪平的时候，这些昆虫的家也随之被铲除了。因此，在现在的大城市中几乎看不到蝴蝶了。

再来看一看凡尔赛宫。以前印象中的凡尔赛宫有被修剪成几何图形的植物以及开阔的视野。如今凡尔赛宫的养护和以前大不相同，用几何图形的预留方式让这些野花野草自然生长（图3-5），为昆虫留下栖身之所，旨在保护自然环境，保护生态系统，保护生物多样性。这个时代的景观设计已经不再偏好对植物的驾驭和配置，而是更多地关注土地上一切生命的和谐共存。

在中国传统园林的工艺案例中也不乏对

Fig 3-4 shows a city green land which can be seen everywhere in France, with one part of lawn trimmed, and the other part untrimmed.

What is the untrimmed lawn for? It's a shelter for insects in the city.

In big cities today, people can hardly see butterflies. Where have they gone? These butterflies mainly spawn on the surface of plant leaves. When gardeners trim lawn neatly, they also destroy the homeland of these insects. As a result, we cannot see butterflies in cities.

Let's have a look at the Palace of Versailles. Previously, it had plants trimmed into geometric shapes and spectacular view. But today, its conservation is totally different. Places of geometric shapes are reserved for the natural growth of wild flowers and grass, also as the shelter for insects, protecting natural environment, ecosystem and biological diversity. Nowadays, landscape architecture has no longer been interested in controlling and arranging plants, but paying more

图 3-5

土地的关注。图3–6是苏州园林对古树根系的保护。这种做法符合树木根系生长的规律，也符合当今海绵城市的设计原则。

图 3–6

attention to the harmonious coexistence of all lives on land (Fig 3-5).

There is no shortage of care for land in cases of traditional Chinese gardens. Fig 3-6 is the protection of root system of old trees in Classical Gardens of Suzhou. It is totally in line with the growth law of tree roots and the design principle of sponge city today.

四

海绵城市的实践与方法

Practice and Methods of Sponge City

俞孔坚指出：我们需要对“海绵城市”概念有更深刻的理解，集中体现在以下几个方面：

第一，完全的生态系统价值观，而非功利主义的、片面的价值观。稍加观察就不难发现，人们对雨水的态度实际上是功利主义的。砖瓦厂的窑工，天天祈祷明天是个大晴天；而经历久旱的农民，则天天到龙王庙里烧香，祈求天降甘霖。“海绵”的哲学是包容，对这种以人类个体利益为中心的雨水价值观提出了挑战，它宣告：天赐雨水都是有其价值的，不仅对某个人或某个物种有价值，对整个生态系统而言都具有天然的价值。人作为这个系统的有机组成部分，是整个生态系统的必然产物和天然的受惠者。所以，每一滴雨水都有它的含义和价值，“海绵城市”珍惜并试图留下每一滴雨水。

第二，就地解决水问题，而非将其“转嫁”给异地。几乎一切现代水利工程的起点

Yu Kongjian points out, we need to understand the concept of “sponge city” more deeply, which is mainly shown in the following aspects:

First, total ecosystem values, but not utilitarianism or one-sided values. If you pay some attention, you will find that people’s attitude towards rainwater is very utilitarian and selfish. Workers in tile factory pray for a sunny day every day, while farmers who experience a long drought pray for heavy rain. The philosophy of “sponge” is inclusive, which challenges the rainwater values that treat people’s personal interest as the center. It declares: rainwater has its natural value, not only to a person or a species, but also to the whole ecosystem. Human beings, as an organic component of this system, is an inevitable product and natural beneficiary of the ecosystem. Therefore, every drop of rain has its meaning and value. “Sponge city” cherishes and tries to save every drop of rain.

Second, solving the water problems on the spot instead of passing it on to another place is the starting and ending point of nearly all the modern water conservancy engineering, but in fact flood levee, remote water transfer etc. all drain flood to downstream and the other riverside, or transfer

和终点，例如防洪大堤和异地调水，都是把洪水排到下游和对岸，或把干旱和水短缺的问题转移到其他地区。“海绵”的哲学是就地调节旱涝，而非转到异地。中国古代的生存智慧是将水作为财富，就地蓄留——无论是来自屋顶的雨水，还是来自山坡的径流，因此有农家天井中的蓄水缸和遍布中国大地的坡塘系统。这种“海绵”景观既是古代先民适应旱涝的智慧，更是地缘社会及邻里关系和谐共生的体现，是几千年来以生命为代价换来的经验和智慧在大地上的烙印。

海绵城市强调将有化为无、将大化为小、将排他化为包容、将集中化为分散、将快化为慢、将刚硬化为柔和。

图4-1展示了河道的硬化给水系带来的

drought or water shortage to other areas. The philosophy of “sponge” is to regulate drought and flood on the spot, but not pass it on to another place. The wisdom for living in ancient China was to treat water as wealth and save it on the spot, no matter it was rainwater from the roof or runoff from the hillside. Thus, there were tanks for water storage in patio of farmhouse and slope-pond systems all over the vast land of China. Such kind of “sponge” landscape was not only the wisdom of ancient ancestors to adapt to drought and flood, but also the embodiment of harmonious coexistence of geopolitical society and neighborhood relations. It was the mark of experience and wisdom on land which was in exchange for the cost of life for thousands of years.

Sponge city focuses on changing existence into non-existence, big into small, exclusion into inclusion, quick into slow and hard into soft.

图 4-1

巨大破坏。河水无法与河底土壤进行有氧交换，微生物系统也无法进行有机交换。

硬化的河道使洪水泄流速度加快，增大了洪水的侵蚀力，也给下游城市带来了更大的水患压力。这既不符合生态水系统组建规律，也不符合海绵城市的设计要求。

图4-2是法国塞纳河河岸整治的状态。河岸采用青石板、石笼和草皮制作成河岸台阶。这是非常具有代表性的海绵河岸，这种河岸透水透气，为城市雨水溢流和地下水渗透创造了便利的条件。在河水上涨时，河岸也可以像海绵一样吸纳一部分洪水。

第三，分散式的民间工程，而非集中式的工程。中国常规的水利工程往往是集中力量办大事的体现，如都江堰水利工程，其在利用自然水过程中的因势利导所体现出的哲

Fig 4-1 shows the huge damage to water system caused by hardening river courses. River water cannot have aerobic exchange with soil on the river bottom, nor can microbial system have organic exchange.

The canalized river course speeds up flood flowing, which increases the erosion force of flood and puts more flooding pressure on downstream cities. This is in line with neither formation laws of ecological water system nor design requirements of sponge city.

Fig 4-2 is the renovation situation of the Seine in France. Riverside steps are made of quartzite, stone cage and turf. This is a very typical sponge riverside. It is permeable and breathable, which creates convenience for rainwater overflow and groundwater permeating in the city. When river water rises, the riverside can also absorb part of flood like sponge.

Third, dispersed non-governmental engineering, not centralized engineering. Usually, conventional water conservancy engineering in China is constructed by the nation or groups. Such as Dujiangyan Dam,

图 4–2

学和工程智慧，使这一工程得以延用至今，福泽整个川西平原。但集中式大工程，如大坝蓄水、跨流域调水、大江大河的防洪大堤、城市的集中排涝管道等，也存在一些不尽如人意的效果。从当代的生态价值观来看，与自然相对抗的工程并不明智，也更加不可持续。而民间分散式的水利工程可持续性更佳。“海绵”的哲学是分散、由千万个细小的单元（细胞）构成一个完整的功能体，将外部力量分解吸纳，消化为无。因此，我们呼吁珍惜和保护民间水利遗产，提倡分散的微型水利工程。这些分散的民间水工设施不仅对自然水过程和水格局不会造成破坏，而且构成了能满足人类生存与发展所需的伟大的国土生态海绵系统。[①]

1. 大尺度：调蓄池区域、民间水利工程、农田、湿地（自然和人工）、森林绿地吸水区域，以及河道缓冲带和河道自然形态

降雨范围并非只限于城市，也包括比城市更大的地域范围。因此，以雨洪管理为主要特征的海绵城市设计不能单纯地依靠城市规划完成，更需要通过更大尺度的土地构建设计、雨洪管理和水系管理完成。因此，包括城市在内的地域降雨给地势较低的城市河

① 俞孔坚等：《海绵城市》，北京：中国建筑工业出版社，2016年版，第12页。

which is still to be used so far and benefits the whole plain in west Sichuan Province due to its philosophy and engineering wisdom of taking advantage of natural water. However, there are a lot of failure cases of centralized big engineering, such as water storage with dam, trans-regional water transfer, flood levees of big rivers, centralized drainage pipes in cities, etc. Seeing from contemporary ecological values, centralized engineering against natural process is not wise, and often unsustainable. Non-governmental dispersed or democratic engineering is usually more sustainable. The philosophy of "sponge" is dispersion. Thousands of small unit cells constitute a complete functional body, which decomposes and absorbs external force to none. Therefore, we call on cherishing and caring about non-governmental water conservancy heritages, and promoting democratic and dispersed miniature water conservancy engineering. The dispersed non-governmental water conservancy engineering and facilities not only won't damage natural water process and water pattern, but also constitute the great land ecological sponge system that satisfies the needs of human existence and development. [①]

1 Large Scale: Region of Regulating Pool, Non-Governmental Water Conservancy Engineering, Farmland, Wetland (Natural and Artificial), Water Absorption Area of Forest and Green Land, Buffer Zone and Natural Form of River

Rainfall not only happens in the city, but in areas

① Yu Kongjian et al., *The Sponge City*, Beijing: China Architecture & Building Press, 2016, P12.

道泄洪带来了主要压力。城市设计无法有效缓解这种压力，其缓解需要在城市之外的土地环境中进行。

1.1 调蓄池区域

在城市之外广阔的自然环境中，从地理的角度可以对地质和土壤状况进行分类，进而确定可以设计区域调蓄池或蓄水池的地块。如，有些区域地下由岩石构成，雨水下渗不畅，不适于做海绵体。那么，这样的地块可以考虑用作调蓄池或蓄水池，收集和储蓄雨水，减少河流的汇水量，降低洪峰形成的概率，从而保护河流下游的城市免受洪灾（图4-3、图4-4）。

1.2 民间水利工程

分散式的民间微型水利工程，如坡塘和水堰（图4-5），至今仍充满活力，受到乡民的细心呵护，并且在海绵城市的治理过程中，发挥着调节雨水汇水流量的重要作用，是缓解城市洪水压力的重要解决办法之一。这些分散式的民间微型水利工程不仅可以协助抗洪，还是合理管理农耕过程的重要手段。[①]

① 俞孔坚等：《海绵城市》，北京：中国建筑工业出版社，2016年版，第12页。

larger than city. Therefore, sponge city design featuring rain and flood control cannot simply be completed by city planning. It should be completed through larger-scale landscape architecture, rain and flood control, and water system management. Therefore, regional rainfall including the city puts main pressure on rivers of lower-lying city. City design cannot relieve these pressure; it needs to be done in the land environment out of city.

1.1 Region of Regulating Pool

In the vast natural environment out of city, geology and soil conditions can be classified from geographical point of view, and furthermore, the block that can be used to design regulating pool or reservoir will be determined. For example, there are rocks under the ground of some places and rainwater cannot permeate smoothly. These places are not suitable to be sponge. Then they can be considered to be made into regulating pool or reservoir to collect and store rainwater, reduce the amount of water flowing into river, and reduce the probability of flood peak, so as to protect downstream cities from flooding (Fig 4-3, 4-4).

1.2 Non-Governmental Water Conservancy Engineering

Dispersed non-governmental miniature water conservancy engineering such as slope pond and weir is still running well and carefully cared by villagers (Fig 4-5). Furthermore, in the management process of sponge city, it plays an important role in regulating the convergence amount of rainwater, and is one of the key solutions of relieving flood pressure of

图 4-3

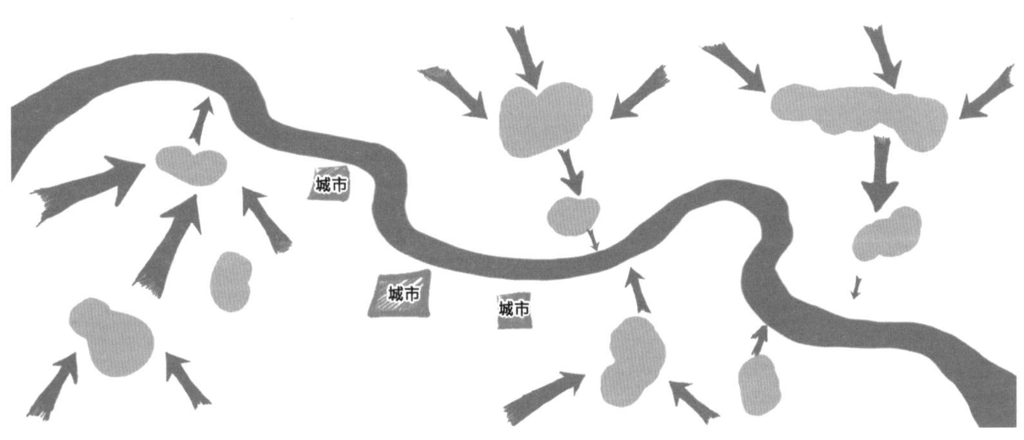

图 4-4

图 4-5

1.3 农田

农田中的生物少于自然草场和森林中的生物，但是农田对于降雨和水系的调节起到了重要作用，是理想的海绵体。因此，生态农业生产性景观配合旅游开发成为近年来的热点。由城市化开发转向对乡村的关注，是景观规划设计划时代的进步。法国20世纪60年代景观规划设计学科的发展就在这种转型下获得了新的契机。设计师们开始关注以往被忽略的地方，如城市与乡村交界处的设计、农田与自然环境的生态过渡、旅游与农业文化体验等。这些项目不但提高了农田的经济效益，而且使农田得到了更加专业的保护和利用，尤其是在今天海绵城市的推进过程中，更是打造了城乡产业互动、经济互惠和文化共享的良好平台。

the city. They can not only assist in fighting against flood, but also manage farming process reasonably.①

1.3 Farmland

Although the biological diversity of farmland is less than that of natural grassland and forest, it is still a key point of regulating rainfall and water system. So it is an ideal sponge. Thus, the productive landscape of ecological agriculture matching with tourism development becomes a hot spot in recent years. Turning from urbanization development to paying attention to villages is also an epoch-making progress of landscape planning and design. The development of landscape planning and design in France in the 1960s got a new chance under such transformation. Designers started to pay attention to places that had been ignored before, such as the design of junction

---

① Yu Kongjian et al., *The Sponge City*, Beijing: China Architecture & Building Press, 2016, P12.

1.4 湿地（自然和人工）

城市之外的自然湿地的保护和扩展有利于海绵城市的降雨和水系调节（图4-6）。这些珍贵的湿地不但可以保护和畜养生态系统中的物种，还可以有效调节大气湿度和温度，是人类生存环境中的“绿肺”。这些城市周边的湿地可以调节城市的湿度和温度，这是对生活在城市中的人们最大的影响。在近代的城市化进程中，城市的范围在扩大，但城市中包括湿地在内的绿地面积却在缩小，尤其是填埋湿地，这将破坏原有湿地及其周边的生态链，同时，生态系统也严重受损。城市周边的湿地是巨大的海绵体，强大的蓄水能力可以在城市旱季和高温天气释放大量的水蒸气，这些水蒸气均衡了城市湿度，同时，也降低了城市温度。这种水气交

of urban and rural areas, the transition from farmland to natural environment, the experience of tourism and agricultural culture, etc. These projects not only enhance the economic efficiency of farmland, but also provide farmland with more professional protection and use. They especially provide a good platform for industry interaction, economic reciprocity and cultural sharing between cities and villages in the process of constructing sponge cities today.

1.4 Wetland (Natural and Artificial)

The protection and expansion of natural wetlands out of city are good for regulating rainfall and water system of sponge city (Fig 4-6). These precious wetlands can not only protect and feed species in the ecosystem, but also effectively regulate atmospheric humidity and temperature. They are “green lungs” in the living environment of human beings. The most important influence of wetlands around these cities on people’s life is on humidity and temperature. In

图4 6

换的过程形成了微物候环境，可以有效缓解城市热岛效应，营造舒适的城市环境。在城市设计中，人工湿地也起到同样的作用。因此，湿地的保护与设计对于海绵城市的功能梳理极为重要。

湿地植被在湿地系统中发挥重要的净化作用，协助城市污水厂尾水再次净化，确保城市水生态的安全。湿地中植物的根系瘤和根系圈通过叶子的光合作用吸附根系周边的重金属等污染物，并将污染物转化或固定在木质素中。常见的湿地植物（如芦苇和香蒲）对氯化钠和重金属等污染物具有很好的吸附功能，并能将无法转化成木质素的重金属污染物锁定在秆茎之中，可以通过焚烧提炼重金属，也可以通过发酵制造沼气。

湿地的净化过程可以大大减轻污水处理的负担，因为在污水处理的过程中，不仅有常规的污染物处理问题，还有很棘手的抗生素净化处理问题。人们在生产、生活和医疗过程中大量使用抗生素，导致人工和自然水体常年含有超量的抗生素，威胁人类健康，湿地的净化过程也为复杂的抗生素净化提供了便利。

在很多实践领域中，“以人为本”是符合客观条件和需求的。但在今天的景观设计（或者称为“土地构建设计”）中，

the process of modern urbanization, the scale of city is expanding while the green area including wetlands inside is shrinking, especially wetland filling, which will destroy the original wetlands and their surrounding ecological chains and seriously damage the ecosystem at the same time. The wetlands around the city are huge sponges with great storage capacity which can release large amounts of water vapor in dry season and hot weather so as to balance the humidity of the city and reduce its temperature as well. The process of water vapor exchange has formed a micro-phenology environment which can effectively alleviate urban heat island effect and create a comfortable urban environment. Artificial wetlands also play the same role in city design. Therefore, protecting and designing wetlands are of vital importance to arrange the function of sponge city.

Wetland vegetation plays an important role in the purification of wetland system. It helps to re-purify the tail water in municipal wastewater treatment plants so as to ensure the safety of urban water ecology. The root tumors and root circle of plants in wetlands adsorb heavy metals and other pollutants around their roots by photosynthesis of leaves and then convert or fix these pollutants into lignin. Common wetland plants such as reeds and cattails can well adsorb sodium chloride, heavy metals and other pollutants and keep those heavy metal pollutants which cannot be converted into lignin locking in the stem so that heavy metals can be extracted by incineration and methane gas can also be produced by fermentation.

The purification process of wetlands can greatly

人已经不再是被服务的主体，也不再是设计的主体。景观设计更关注的是所有生命（包括植物在内）在生存环境中如何和谐共存。因此，在这种可能性的设计中，“人”仅仅是一个应该介入的元素，而不是“中心”。

湿地中的鸟类经常成为开发商和设计师关注的“亮点”，因此有很多的观鸟设计。遗憾的是大多数的设计师在勾画美丽的观景台和观景廊道时，只考虑人的便利和对人的服务，如将停车场设计在距离观鸟地最近的湖畔和水边，而没有考虑到各种鸟类的安全警戒距离及鸟类的独立生存环境（图 4-7）。结果，人们还没有到达观景台，鸟

reduce the burden of sewage treatment because there are not only problems about conventional pollutant treatment, but also tough treatment of antibiotics. The extensive use of antibiotics in production, life and medical process results in artificial and natural water containing excessive antibiotics year in and year out, which threatens human health. Therefore, the purification process of wetland also provides convenience for complex purification of antibiotics.

In many practical fields, “human oriented” is in conformity with objective conditions and requirements. However, for landscape design (or known as “landscape architecture”) at present, people are no longer the main part of service, nor design. Landscape design is more concerned with the possibility of all life (including plants) achieving harmonious coexistence in the living environment. Therefore, in the design of such possibility, human is

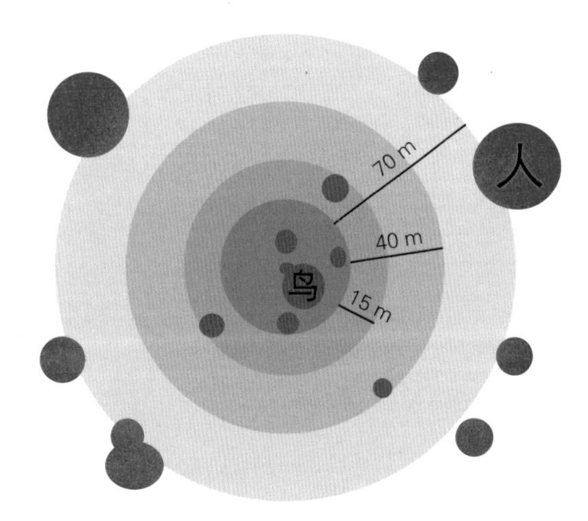

图 4-7

儿就已经飞走了。另外，停车场中的汽车漏油通过雨水的淋蚀渗透污染了水域，含铅过高的多种重金属污染物会毒死水中的鱼虾，鸟儿的食物来源被切断，游客还能够观赏到鸟儿吗？我们应该设计的是人与鸟共生的环境，而不是让鸟儿成为为人类服务的工具和对象。景观设计研究的是一切生命和谐共存的关系，并且使人类在这种和谐中受益和发展。

1.5 森林绿地吸水区域

城市外的森林绿地是重要的海绵体，它们迅速吸收雨水，并补给地下含水层。地球上最大的水库就是地下含水层，它可以通过自然的水循环净化水质、释放氧气和分解二氧化碳。那些被城市外的森林绿地大量吸收的降雨，有效地缓解了城市洪灾；雨水通过渗流作用汇入地下含水层和河道，有效地争取了错峰时间。

1.6 河道缓冲带和河道自然形态

河道缓冲带是河道生命机能的重要器官，是天然的海绵体。人为开发占用和硬化了河道缓冲带，破坏了河道缓冲带的海绵体功能，这是增加洪水威胁的主要原因。

另外，为了通航方便，河道截弯取直（往往就是河道缓冲带）导致泄洪威胁增

only an “element” which should be involved but not the “center”.

Birds at wetland often become the “bright spot” focused by developers and designers and thus a lot of bird-watching designs are produced. Unfortunately, most of the designers only consider the convenience of people and services for people by designing parking lots besides the lake and water which are the nearest to birds and have not taken the safe vigilance distance of various birds and their independent living environment into consideration during delineating the beautiful viewing platform and viewing gallery (Fig 4-7). As a result, birds fly away before people reach the view platform. In addition, the leakage of car oil pollutes water through the erosion and penetration of rain, and a variety of heavy metal pollutants containing too much lead poison small fish and shrimp, without which birds cannot live. The food source of birds has been cut off; then how can tourists see birds? We should design an environment where both humans and birds could exist rather than taking birds as the tools and objects for human services. Landscape design is exploring the harmonious coexistence of all lives and also enables human beings to benefit and develop in such harmony.

1.5 Water Absorption Area of Forest and Green Land

Forest and green land out of city are important sponges, which quickly absorb rainwater and supply to underground aquifer. The largest reservoir on earth is underground aquifer, which can purify water, release oxygen and decompose carbon dioxide through natural water cycle. Rainwater, which is

加，洪峰通过速度加快，下游危害更大，直接破坏了生态系统；通过水构建的微生物系统和生物栖息地被摧毁；河床土壤也无法通过水的渗透进行呼吸，地表水与地下水无法进行有益交换。

图4-8是法国亚眠市河道改造的效果。河道转弯处设有低矮的木质挡板来抵抗洪水侵蚀时造成的水土流失，也保护了水生动物栖息地。这些挡板之间的缺口可供水生动物自由进出。

图4-9是莱茵河穿越德国杜塞尔多夫市中心的改造效果。莱茵河右岸是商业中心和办公区，莱茵河左岸留有约1.5 km的自然滩涂。这些预留的自然滩涂正好是莱茵河穿越

largely absorbed by forest and green land out of the city, effectively reduces flood in the city. Rainwater permeating into underground aquifer and river provides more time for flood peak.

1.6 Buffer Zone and Natural Form of River

Buffer zone is an important organ for river to keep life function, and also a natural sponge. Being developed and occupied by human beings leads to its hardening, which is the main cause of flood.

In addition, curving river course (always buffer zone) is cut to be straight for the convenience of navigation. It increases the threat of flood draining, with flood peak passing quicker and downstream bearing more harm, and it also destroys ecosystem directly. Microbial systems and biological habitats relying on water are also destroyed. Riverbed soil cannot

图 4-8

图 4–9

市区的河道缓冲带。左岸的河道缓冲带通过水流的漩涡沉积了大量的泥沙，泥沙中的丰富营养使左岸水草生长得尤为丰茂。照片中的白点是小羊羔，每个星期一的上午9点~12点，牧民带着这些小羊羔过来吃草，小羊羔们把草吃得整整齐齐，比园丁修剪得还要好，这项设计既节省了人力，同时也创造了城市中的乡村景观。该设计的出现甚至还使杜塞尔多夫的抑郁症患者的患病率明显下降。

图4-10a中的法国索米埃市洪水公园是河道水系整治的成功案例。该公园地处维杜尔河的河道缓冲带，凶猛的洪水不断冲击着罗马时期的千年古桥。该公园的设计保护了古桥免受洪水袭击，也发挥了河道缓冲带拦截洪水冲积物和减缓水流速度的作用。

索米埃市地处平原的洼地，穿越索米埃市的维杜尔河每年都有洪水泛滥，一段河道

breathe by the permeation of water, and favorable exchange between surface water and groundwater becomes impossible.

Fig 4-8 shows the effect of river course renovation in Amiens, France. Low wooden boards are set up in the turning places of river to resist water loss and soil erosion when flood comes, and also to protect habitats of aquatic animals. The gaps between these boards enable free access of aquatic animals.

Fig 4-9 shows the renovation effect of the Rhine flowing through center of Dusseldorf. The right side of the Rhine is business center and office zone and on the left side, there is a natural beach of about 1.5 km reserved. The reserved beach is just the buffer zone of the Rhine flowing through the urban area. Due to the whirlpool of water, buffer zone on the left side carries and deposits a lot of silt. The rich nutrition in the silt makes water plants flourishing. White points in the picture are lambs. Each Monday from 9 to 12 a.m., herdsmen take them here to eat grass, which makes grass even neater than gardeners' trimming. This design not only saves manpower, but also creates

随着时间的推移和洪水多次的冲击，泥沙沉积，河流改道，原有河道在非洪水季节处于干涸状态。多年淤积的河泥使这片干涸的河床土壤肥沃，农民在汛期前后种植多种农作物，但是河水在汛期经常漫过河床，冲刷着河岸的树林和农田并携带大量的残枝枯木，这些残枝枯木极易造成各种危害。[①]

country landscape in the city. What's more, it makes a significantly decrease in the rate of depression in Dusseldorf.

Fig 4-10a shows that the flood park in Somieres, France is a successful case of river water system renovation. This park locates on the buffer zone of Le Vidourle River. The ferocious flood keeps attacking the millennium old bridge of Roman times. The design of this park protects the old bridge from being

图 4-10a

① 安建国，方晓灵：《法国景观设计思想与教育》，北京：高等教育出版社，2012年版，第69页。

索米埃市洪水公园的设计需要解决三个问题：第一，加强河床和河基的保护，防止水土流失；第二，防止洪水携带物（残枝枯木、冲垮的房屋和汽车等）破坏索米埃市，减少洪水流量，减缓洪水流速；第三，旧河道及其周边土地的农业利用，以及如何发挥其社会公益价值。

公园位于河道转弯处，在洪水季节，河床和河岸遭到300 m^3/s的洪水冲击破坏。为了加固修整河岸，设计师调整了河岸坡度，并砍伐了一部分由于地形改变而无法保存的树木。设计师利用砍伐的树木制作了5 m长的双排树桩，固定在河岸常规水位线上，两排树桩之间由捆绑好的树枝填充，并用铁丝固定。调整后的河岸坡度为30°，河岸泥土被压实后铺上一层有机植物材质制作的护坦（用来抵御洪水冲刷），这些护坦用铁丝和铁钉固定在用来保护河岸的树桩上，长1 m、直径约5 cm的小树桩以3 m的间距排列，点状分布固定在护坦上，然后在小树桩上缠绕铁丝形成铁丝网再次加固护坦，最后在护坦上掏孔栽种树苗，这些由植物原料制成的护坦随时间的推移将腐化成泥土，护坦在小树长大前保护和固定树根土壤，并减少杂草滋生。河流在穿越索米埃市罗马古桥之前要经过一座钢筋混凝土桥——索米埃市新桥，这座桥是连接两岸最重要的交通枢纽，但该桥每年

attacked by flood and also enables the buffer zone to play a role in stopping the objects brought by flood and slowing down the speed of water flow.

Somieres is located in the low-lying plain. The Le Vidourle River which flows through Somieres floods every year. As time goes by and flood hits on, mud and sand deposit, so the river has to change its course. The original course is dry in non-flood season. Mud that has deposited for many years makes the soil of this dry course fertile, and farmers plant many kinds of crops before and after the flood season. However, river water often overflows the riverbeds, washing forest and farmland on the bank and carrying lots of dead branches that can cause various harm.[①]

The design of flood park in Somieres needs to solve 3 problems. First, strengthen the protection of river bed and base to prevent water and soil erosion. Second, prevent things carried by flood (dead branches, collapsed houses and cars, etc.) from destroying Somieres, and reduce flood flow and slow down its speed. Third, use the old river course and land surrounding it for agriculture, and how to show its social public welfare value.

The park is located on the turning place of the river course. In flood season, the river bed and bank are hit by 300 m^3 of flood per second. To reinforce and repair river bank, designers adjusted the bank slope and cut down part of the trees that could not remain due to terrain change. Designers made these trees into 5-meter double-row tree stumps and fixed them on the conventional water level line of river bank.

① An Jianguo and Fang Xiaoling, *Thought and Education of Landscape Architecture in France*, Beijing: Higher Education Press, 2012, P69.

都受到洪水的冲击和侵蚀，尤其是从上游冲刷下来的残枝枯木和汽车经常堵截在石墩之间，使河流无法顺利泄洪，很可能会冲垮桥身。另外，被堵截的洪水将淹没该桥上游大部分的农田和农舍，设计师对此深入研究后，通过各种方式充分利用土地资源，变灾害为财富。

除了河岸之外，洪水公园的设计有六个组成部分（图4-10b）：

第一部分是临近洪水冲击的主要区域，在第二、第三部分有六处取水钻井，所以第一部分区域使用了有机植物材料做护坦，并在河滩上种植科本植物，目的是防止水土流失和洪水带来的大量泥沙渗入地下，污染地

The bound branches fill up the spaces between tree stumps and are fixed by iron wire. The adjusted bank slope is 30 degrees. Soil of the bank is relatively compact, and then covered by a layer of protectors made of organic plant materials (to resist flood washing). These protectors are fixed on the tree stumps, which are used to protect river bank, by iron wire and nails. Tree stumps of one-meter long, about 5-centimeter diameter are dotted and fixed on the protectors with 3-meter spacing, and wind the tree stumps with iron wire to form nets to reinforce the protectors again. At last, dig holes on the protectors and plant saplings. These protectors made of plant materials will corrupt into soil as time goes by. They protect and fix root soil and reduce the growth of weeds before trees grow up. Before flowing through the old bridge of Roman times in Somieres, the river first passes a reinforced concrete bridge, which is

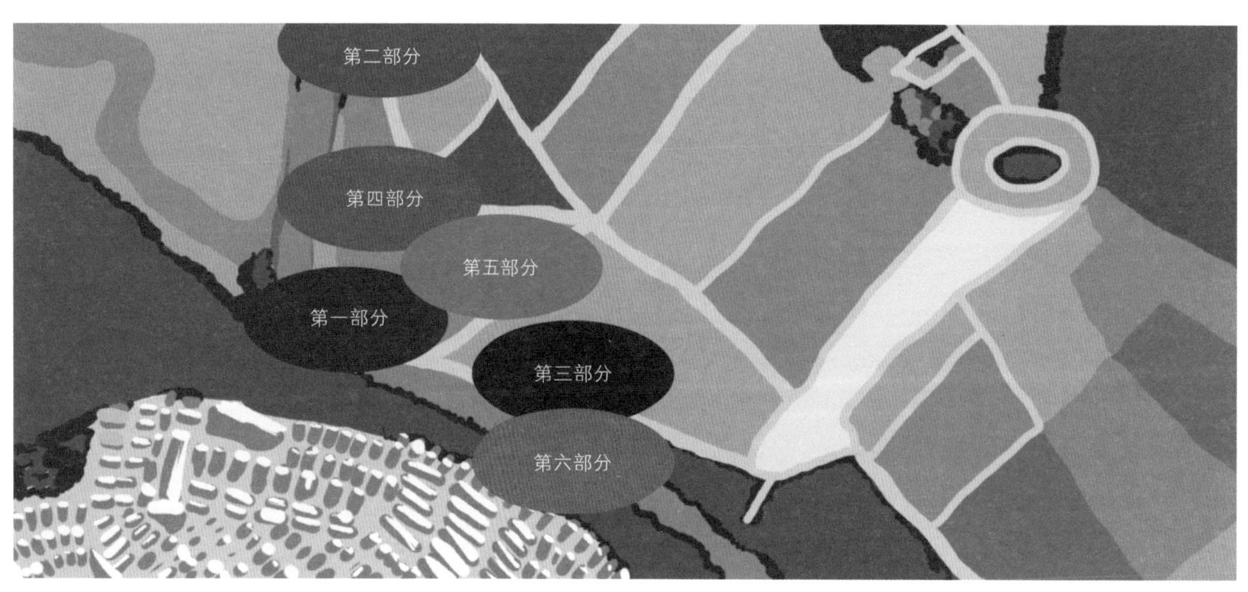

图 4-10b

下含水层和取水源。

第二、第三部分种有橡树、枫树、榆树、白蜡树、胡桃树等树苗，在树苗之间根据季节种植农作物，每年洪水带来的淤泥中含有丰富的养料，所以这里的农作物和树木（可以用来制作家具）长势喜人，这里既是农田又是天然苗圃。

第四部分在铺设护坦后加上5 cm厚的石子，这些石子可以固定护坦，也可以提供泥沙沉积的空间，在洪水过后的第二年春天，人们可以在这里欣赏野花野草之美。

第五部分是公园设计的核心部分（图4-11a），由一排直径为60 cm，前后间隔1 m，均匀错位排列的铁柱组成，柱高10 m，地下6 m、地上4 m（图4-11b），柱头探出的鸟嘴部分用于拦截洪水带来的枯木，并防止由于水流过大导致枯木浮过铁柱撞击桥身。铁柱采用自然的铁锈红色，柱身与地表

the most important transportation hub connecting the two banks, but it is attacked and eroded by flood every year, especially dead branches and cars from upstream, which block the space between stone piers. As a result, flood cannot be drained smoothly and the bridge might be washed down. In addition, the blocked flood will submerge most farmland and farmhouses upstream of the bridge. Designers deeply studied this and made full use of the resources taken by land and flood in various ways, and changed disaster into wealth.

Besides river banks, there are 6 components in the design of flood park (Fig 4-10b).

The first part is the main area that is near to the flood's attack. This area uses protectors made of organic plant materials, and plants are planted on beach to prevent a large amount of mud and sand, which are brought by flood and water and soil erosion, from permeating into the ground, otherwise they may pollute underground aquifer and water source because there are 6 water drillings in the second and the third area.

Oak, maple, elm, ash trees, walnut trees and other

图 4-11a

图 4-11b

水平交界处浇注水泥加固，铁柱边的护坦由金属连接件浇筑固定在水泥基座中。这种做法可以防止洪水冲击导致的水泥周围的土壤流失，不影响水泥基座的稳固性。铁柱阻截了洪水中的树枝枯木和汽车，并大大减缓了水流速度。洪水在通过这一排铁柱后的20 m处，又坠入农田排水用的沟渠，并形成水流回旋，从而又一次降低了水流速度。这样，相对平静的洪水通过第三区域的树林后，携带的泥沙沉积在树林中，洪水经过铁柱和树林的阻截后安全平静地流过索米埃市。这些铁柱不仅具有抗洪功能，在非洪水季节，也发挥着独特的艺术和社会价值。

人们漫步在这些铁柱之间，欣赏从不同角度呈现出的美景。人们寻找着铁柱与肢体之间的比例差异，有时仰视、有时俯视、有时远眺、有时后退、有时加快脚步、有时躲在铁柱之间、有时寻找透视效果……这些铁柱给人们带来了肢体与空间的对话，带来了无限的想象，其次序性给人们带来了强烈的视觉冲击力，人们在不同距离能感受到不同的空间效果。这些看起来笔直的铁柱事实上是微微倾斜的，工程师做了严谨的洪水水流方向和冲击力系数分析，让这种弧线形和前后错位的铁柱可以均匀地承担洪水的冲击力。

此外，这些柱列在视觉上可能给人带来

saplings are planted in the second and third area, and crops are planted among saplings according to different seasons. Since there are rich nutrients in the silt brought by flood every year, crops and trees (which can be used to make furniture) here grow quite well. This place is both farmland and natural nursery.

In the fourth area, 5-centimeter-thick stones are added on protectors. These stones can fix protectors and effectively provide space for mud and sand deposition. In next spring after flood, people can enjoy beautiful scenery of wild grass and flowers here.

The fifth part is the core of park design (Fig 4-11a). It consists of a row of 60-centimeter-diameter iron pillars that are evenly distributed, with an interval of one meter. The pillar is 10-meter high, 6 meters underground and 4 meters on the ground (Fig 4-11b). The beak part out of the pillar is used to block dead branches brought by flood, and prevent them from floating over the pillars to hit the bridge when water flow is too large. The colour of iron pillars is natural rust red. The horizontal junction of pillars and surface is reinforced by cement. Protectors besides pillars are poured and fixed in the cement base by metal connectors. This can prevent soil around cement from being washed away and avoid influencing the stability of cement base when flood comes. Iron pillars block dead branches and cars in flood and largely slow down the water speed. Flood flows down to the farmland drainage ditch after passing this pillar rows for 20 meters, and forms water roundabout, which slows down the water speed again. Thus, when the relatively calm flood flows through the forest of the third part, mud and sand carried in the flood deposit in the forest. After being blocked by iron pillars and

错觉，使游客下意识地改变观察和体验方式，更具空间审美价值，并积极地调动着周边的空间元素，其微妙变化形成了不同空间尺度的感知效果。这些原本不被青睐的钢筋混凝土成了该设计最精彩的部分。因为除了实用意义外，它们还在不断创造参与和分享空间的可能。

公园的第六部分是一个人造的“避风港”（图4-12），设计师细心地为动物保留了这一处免受洪水侵袭的区域。避风港上游有一个小闸口，引河水缓缓流入，来保持港内清净的水质和适度的含氧量。避风港常规水位线处都采用双排树桩加固并由树枝填

forest, the flood will safely and quietly flow through Somieres. These iron pillars can not only control the flood, but also have unique artistic and social values when there is no flood.

Wandering among these iron pillars, and enjoying the scenery from different angles, people look for the proportion differences between iron pillars and limbs; they sometimes look up, sometimes look down, sometimes look at a distance, sometimes go backward, sometimes walk faster, sometimes hide among pillars, and sometimes look for perspective... These iron pillars bring people dialogues between limbs and space, as well as infinite imagination. The sequence of these iron pillars brings strong visual impact. At different distances, we can feel different space effects. These iron pillars that seem to be straight are actually slightly slanting. Engineers did

图 4-12

充，为多种水生和半水生动物营造出理想栖息地，这是保护该地区生态平衡的重要措施。

2. 中尺度：利用错峰时间，快速使用地下管网和地表溢流，排出城市内涝

城市地下排水管网在何时才能发挥其最有价值的效应？答案是在暴雨短时间内连续骤降时。此时，城市管网如果没有能力及时排出城市内涝，等到洪峰到来时，河水上涨回灌，内涝就无法排出，加之持续的降雨就形成了城市水灾。

3. 小尺度：城市区域内的技术设计要点“渗—滞—蓄—净—用—排”

通过这六种技术在城市尺度内吸收雨水资源，利用土壤和植被自然净化水质，再利用雨水资源，涵养并补充地下水，改善城市微物候，调节城市湿度和温度，减缓城市热岛效应。

3.1 渗

渗的技术要点是提供一切可以让雨水就地自然下渗的可能和机会，力求将85%左右的雨水在原地下渗吸收或滞留，涵养地下水。以往城市内约85%的降雨通过地表径流和地下管道排走，雨水资源没有得到合理利

a rigorous analysis of flood direction and impact coefficient, and such kind of arc-shaped arrangement enables iron pillars to bear the impact of the flood equally.

In addition, these pillar rows may form illusion visually, which makes tourists change observation and experience modes subconsciously, thus improving its space aesthetic value and actively mobilizing space elements in the surrounding. The slight changes of pillar rows form perceptual effects of different spatial scales. The reinforced concrete which originally was not favored by the landscape has become the most wonderful part of this design because besides practical significance, it creates possibilities of participation and sharing constantly.

The sixth part of the park is an artificial “shelter” (Fig 4-12). Designers carefully reserved this area that cannot be attacked by flood for animals. In the upstream of the shelter, there is a small gate, through which the river slowly flows in, to keep water clean and leave moderate oxygen content in the harbor. The section of the conventional water level line of the shelter is reinforced by double-row tree stumps and filled with branches. This creates an ideal habitat for various aquatic and semi-aquatic animals, and is an important measure to protect the ecological balance of this region.

2 Middle Scale: Quickly Drain Water Logging Through Underground Pipelines and Surface Overflow in Peak Staggering Time

When can urban underground drainage network play its most valuable role? It is when rainstorm

用，反而给江河湖泊增大了汇水压力，甚至导致并加剧了洪灾。

渗的技术可以通过以下几种设计形式表现：

（1）城市路面渗透铺装；

（2）雨水花园或生物滞留措施；

（3）屋顶绿化；

（4）植草沟。

法国凡尔赛宫从宫殿主建筑到十字湖的路面设计都成功地疏导了雨水，减少了水土流失，且保证雨水下渗顺畅。通过图 4-13，我们能够看到路面的“之”字形有微微隆起的部分。这些微微隆起的土包引导着雨水缓慢地沿着“之”字形流动，并且每次流到转弯处都自然地汇流到地下排水管网。这样设计的“之”字形汇水路面，减少了雨水径流对砂石路面的侵蚀，而且砂石路面渗水效果也非常好。这是一个成功的海绵城市道路设计案例。

历经了三个多世纪的法国皇家园林工艺，对今天海绵城市的设计仍具有重要的启发意义。例如，巴塞罗那奥林匹克公园路面设计正是受到凡尔赛宫路面设计的启发。如图4-14所示，巴塞罗那奥林匹克公园道路微微隆起的部分将雨水疏导到一侧的石笼中。雨水通过石笼自然下渗，浇灌石笼另一侧的植被。石笼内使用的是石灰石，多年的雨水

continuously falls in a short time. At that time, if urban network is not able to drain water logging in time, it cannot be drained when flood peak comes, river rises and flows backwards, plus continuous rainfall, thus causing water disaster in the city.

3 Small Scale: Key Points of Technical Design Within Urban Area: Permeating-Delaying-Storing-Purifying-Using-Draining

By the 6 technologies in the city, we can absorb rainwater resources; make use of soil and vegetation to naturally purify water; reuse rainwater resources; conserve and supply groundwater; improve urban micro-phenology; adjust urban humidity and temperature; and reduce the influence of urban heat island effect.

3.1 Permeating

The technical point of permeating is to design all the possibilities and chances for rainwater to naturally permeate, with about 85% of rainwater permeating, absorbing or staying on the spot. In the past, 85% of rainwater in the city was drained through surface runoff and underground pipelines, which was not reasonably used. On the contrary, rainwater adds convergence pressure on rivers and lakes, and even leads to and aggravates flood.

The technology of permeating can be performed in several design forms:

(1) Permeating pavement on urban roads;

(2) Rain garden or biological delaying measures;

(3) Green roof;

(4) Grass ditch.

图 4-13

图 4–14

淋蚀和氧化作用使石灰石结块。这种结构既结实又透水。

图4–15是巴黎雷诺新区的人行道设计，除了自行车道和机动车道之外，人行道已经不再使用硬质石板铺装，而使用配有钙化粉末的细小砾石，通过使用者的踩踏，自然夯实，而且透水效果极佳。这样既节省了人力、物力，又达到了理想的海绵路面效果。

图4–16是法国雷恩市生态社区人行通道设计，通过不同材料的应用，自然地表现出该通道对于不同人流所发挥的不同作用。该人行道宽约5 m（也是供汽车通过的应急通道），中间铺设2 m宽的实心水泥方砖，两侧各铺设1.5 m宽的空心水泥方砖，空心砖的空隙中长满野草，人们从一定的角度可以看到两侧1.5 m宽的、绿油油的草坪。[①]

From the main building of the palace to the road of cross lake of Versailles, the design successfully dredges rainwater, reduces water and soil erosion and makes rainwater permeate smoothly. In Fig 4-13, we can see that there are slight “zigzag” bulges on the road, which lead rainwater to flow in a “zigzag” way slowly and converge to underground drainage network every time it flows to the turning. Such a design of “zigzag” road for water convergence reduces rainwater runoff’s erosion on gravel road; meanwhile, the seepage effect of gravel road is very good. This is a successful case of sponge city road design.

French royal garden technologies that lasted for more than 3 centuries are still of significance to inspire sponge city today. For example, the road design of Barcelona Olympic Park is inspired by that of Versailles. As shown in Fig 4-14, the slight rise on the road of Barcelona Olympic Park leads rainwater into the stone cages on one side. Rainwater naturally permeates through stone cages and waters vegetation

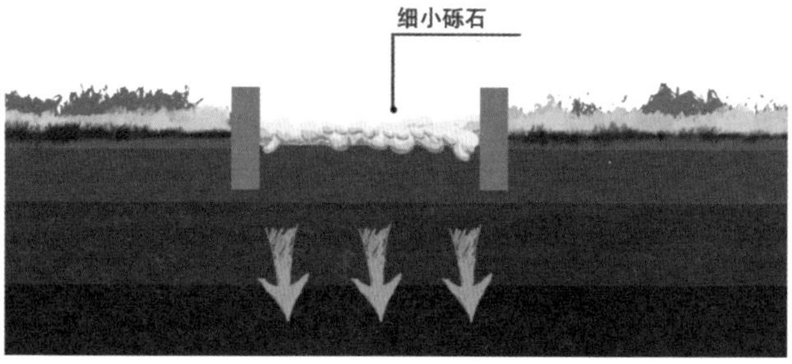

图 4–15

细小砾石

① 安建国，方晓灵：《法国景观设计思想与教育》，北京：高等教育出版社，2012年版，第107页。

图 4–16

从图4–16中，我们能够清晰地看到在5 m宽的道路中心铺设实心水泥方砖的路段最为显眼，由于材料上的差异，这一部分在人流较小时使用率最高。在人流过大时，人们可以自由地行走于空心水泥方砖之上，不影响草的生长，这样做非常有效地利用了土地空间，既解决了不同人流量的通行问题，也扩大了社区的绿化面积，最大限度地减少了水泥对地表土壤的封堵，增加了土壤与大气的氧气交换量，提高了土壤含水率，从而减小了暴雨季节雨水给河道带来的排洪压力。

图4–17是法国北部20世纪60年代废弃的

on the other side. Limestone is used in stone cages. After several years of erosion and oxidation of rainwater, there will be limestone agglomeration, which is strong and permeable.

Fig 4-15 is the sidewalk design of Renault New Area of Paris. Except bike lanes and motorways, hard slates are not used to pave sidewalks. Instead, gravel with calcified powders is used. They are naturally compacted by users' steps, and have excellent permeable effect. This not only saves human and material resources, but also achieves ideal effect of sponge pavement.

Fig 4-16 is the sidewalk design of ecological community in Rennes, France. The application of different materials naturally reflects its different

图 4-17

煤矿区经过了约40年的生态恢复后的景象，这里有很多野兔、刺猬、水獭、鼹鼠和多种珍稀鸟类，甚至还发现了从附近森林里跑来安家的野猪和麋鹿，我们已无法想象这里曾经是寸草不生的矿区。[②]

对于这种高难度的矿区生态环境改造，景观设计师采取了最少人为干预，让自然以自我管理、自我恢复的方式来构建生态环境。如图4-18所示，山坡上挖有深约60 cm的小沟壑，回填10 cm ~15 cm厚的黏土和腐殖土，这些黏土和腐殖土将通过山坡汇流下来的雨水滞留在小沟壑当中，设计师播撒了芦苇之类的草种。此后的生态环境恢复就是大自然自我管理的过程，风会吹来种子，水会带来种子，鸟会衔来种子，在适当的湿度和

② 安建国，方晓灵：《法国景观设计思想与教育》，北京：高等教育出版社，2012年版，第81页。

functions with different flows of people. The sidewalk is about 5-meter-wide (which is also an emergency exit of cars). 2-meter-wide solid cement bricks are paved in the center, and 1.5-meter-wide hollow cement bricks are paved on both sides. Spaces in the hollow bricks are filled with weeds, and within a certain perspective view, we can see 1.5-meter-wide green lawns on both sides. [①]

From the picture, it is clear to see that on the 5-meter-wide road, the central part with solid cement bricks is the most obvious part of the road. Due to the difference of materials, this part is most used when there is a smaller flow of people. When the flow of people is too large, people can freely walk on hollow cement bricks without influencing the growth of grass. This method uses land space very effectively. It solves the passing problem of different amounts of people, enlarges green area of the community, reduces the cement's block on surface soil to the

① An Jianguo and Fang Xiaoling, *Thought and Education of Landscape Architecture in France*, Beijing: Higher Education Press, 2012, P107.

图 4-18

温度下矿渣山开始出现植被覆盖的景象。经过40年，这个矿渣山已经成为人们最喜爱的公园绿地。

图4-19是矿渣山山脚下的人工湿地。在湿地中有二十几处月牙形的装置，这些装置皆由矿渣山砍伐下来的枯木、枯枝以及收集的落叶和细枝制成，由设计师重新安置在人工湿地里。这些月牙形的装置可以为湿地中的小动物提供栖身之所，同时，岸边观景台附近的月牙形装置也有防止水土流失的作用。在这里，人们会经常看到各种野鸭和天鹅，还有探出头来望着你的水獭。这是1960年工业革命结束以后，法国最为杰出的

maximum extent, increases oxygen exchange between soil and atmosphere and improves the moisture content of soil. Thus, it reduces flooding pressure on river brought by rainwater in rainstorm season.

Fig 4-17 is an abandoned coal mine area in northern France in the 1960s. After about 40 years of ecological restoration, there are a lot of hares, hedgehogs, otters, moles and various rare birds living here. Even wild boars and elks from nearby forest are found to settle here. We cannot imagine that here once was a barren mine area.[2]

For the difficult transformation of ecological

② An Jianguo and Fang Xiaoling, *Thought and Education of Landscape Architecture in France*, Beijing: Higher Education Press, 2012, P81.

图 4-19

海绵城市设计案例之一。在城市硬化的空间中，寻找雨水下渗的机会。图中的石钉和灰色砖块可以使雨水迅速下渗，且在它们之间设有小面积的绿化带，便于雨水汇流下渗和滋养植物。

图4-20是法国巴黎阿拉伯世界文化中心，设计师重新将植物种植在已经硬化的广场路面上，这不仅为生硬的石板广场增添了绿色，也为雨水在广场的下渗创造了便利条件。

图4-21拍摄于从法国到瑞士的高速公路，这段高速公路地处阿尔卑斯山山区，海拔在1500 m以上。这里常年积雪，设计师利用这一特点在高速公路斜坡上设置了倾斜

environment in mining area, landscape designers intervene to the minimum degree and allow nature to form ecological environment in the way of self-management and self-recovery. As shown in Fig 4-18, small gullies of about 60-centimeter-deep were dug on the hillside and filled by 10 to 15 centimeters of clay and humus soil, which will reserve rainwater from the hillside in small gullies. Designers sowed grass seeds such as reed. After that, the restoration of ecological environment was a self-management process of nature. Wind, water as well as birds would take seeds here, and in appropriate humidity and temperature, the slag mountain would be covered with vegetation. After 40 years, the slag mountain has become people's favourite park green land.

Fig 4-19 is the artificial wetland at the foot of the slag mountain. In the wetland, there are over 20 crescent-shaped devices like this, which are made of

图 4–20

图 4–21

45° 的漏网。漏网的左侧种植了很多灌木，漏网的右侧用于阻隔、蓄积雪。为何选择倾斜45° 来安置漏网呢？一是在积雪过多时，这样可以避免积雪压坏漏网，让多余的积雪通过重力作用自然滑落到高速公路两边。二是每年春天被蓄积在网格右侧的积雪可以持续缓慢地融化，慢慢地浇灌漏网左侧的灌木丛。这项设计不但为山地绿化提供了持续的水源，而且可以节省大量的人力和资金投入。

图4-22是法国北部图尔昆市空地艺术中心的屋顶花园，是景观设计师和艺术家共同创作的作品，场地位于一个有着近150年历史的废弃纺织厂的屋顶。该作品极为简洁，

dead wood and branches cut from the slag mountain, as well as fallen leaves and thin branches collected. Designers rearranged them in the artificial wetland. These crescent-shaped devices provide habitats for small animals in wetland. The ones near the bank viewing platform in the above picture also have the function of preventing water and soil erosion. Here you can often see various wild ducks and swans. Otters also often look at you with head out. This is one of the most successful cases of sponge city design in France after the Industrial Revolution ended in 1960. Chances are sought to permeate rainwater into ground in the hardened space of the city. Stone nails and gray bricks in the picture enable rainwater to permeate quickly, and small areas of green belts were designed among them, which is convenient for the convergence and permeating of rainwater, as well

图 4-22

既不需要景观设计师进行绿色空间设计和植物配植，也不需要艺术家进行造型设计，设计师们只需要把纺织厂屋顶历经150年所沉积下来的尘土收集起来，重新铺设在加高的厂房屋顶回廊。设计师没有种植任何植物，而是使屋顶尘土中的养分和本土植物的种子通过风、雨、鸟等传播媒介安家落户到这里。[③]

尘土重新铺设后的第三年，屋顶已经郁郁葱葱了，有300多种植物在此安家落户，植物学家们还发现了一些来自中国、印度和非洲的不属于本地区的物种。植物学家解释道：在工业革命时期，这里曾是法国著名的纺织工业基地，有来自世界各地的棉麻原料。这些植物种子被携带在棉麻原料中传入这片土地，有些种子适应了这里的土壤和气候，于是就在此繁衍生息了。

植物学家给屋顶花园里的物种贴标签进行说明，这里不仅给人们带来超越三维空间的历史回忆，也成为孩子们认识植物的花园和教育基地。该作品改变了人们对历史的纪念方式和对生活的评价方式。通过该作品，我们认识到：不仅博物馆的展品可以有纪念意义，不仅一件经典雕塑可以再现和追溯历史文化，景观设计也可以用其独特的表现方

③ 安建国，方晓灵：《法国景观设计思想与教育》，北京：高等教育出版社，2012年版，第115页。

as nourishing plants.

Fig 4-20 is Arab World Institute in Paris, France. Bringing plants back to the hardened square pavement not only adds green to the hardened slate square, but also provides convenience for the permeating of rainwater in the square.

Fig 4-21 is taken on the highway from France to Switzerland. This section of highway is located in the Alps, 1,500 meters above sea level. There is snow all the year round here. Designers took advantage of this and installed nets at a 45-degree angle on highway slopes. On the left side of these nets, many bushes are planted, and the right side is for blocking snow. Why do designers choose 45 degrees to install nets? It's to prevent snow from damaging nets when there is too much snow, and make extra snow slide to both sides of highway naturally through gravity effect. Every spring, snow stored in the right side of nets melts continuously and slowly, watering bushes in the left side slowly. This design provides continuous water source for mountain greening, and saves a large number of human and financial resources.

Fig 4-22 is the roof garden of Open Space Art Center in Tourcoing, northern France, a joint work of landscape designers and artists on an abandoned roof of a textile mill, which has a history of nearly 150 years. It is extremely simple, with neither landscape designers' green space design and plant configuration nor artists' modeling design. The only work of designers was to collect dust on the roof of the textile mill with the history of about 150 years, and repave it on the heightened roof corridor of the mill. Designers planted nothing. Nutrients in the roof dust and seeds of local plants settling here were brought by medium

式对历史文化和现代生活进行诠释。人们的生活在改变，人们对社会的认识和价值取向在改变，一件杰出的作品是无法预见的，甚至是不被觉察的，它的完成需要时间和自然的共同参与。

通过海绵城市的设计，我们不仅要提升自然生态水平，而且要提升社会人文生态水平，让中国社会在海绵城市的建设过程当中，走出具有物质文明和精神文明特色的道路。

3.2 滞

滞的技术要点是设计可以让雨水滞留下来的措施，通过下渗净化水质，补充地下水，延长雨水汇流到河道的时间，从而赢得“错峰”时间，延缓洪峰的到来，降低洪峰流量，减轻洪水威胁。

滞的技术主要是分散式滞留措施：

（1）植被缓冲带、下凹式绿地渗透池、湿地；

（2）雨水截留与收集利用；

（3）街边生态滞留区；

（4）雨水花园和景观水体。

法国凡尔赛宫园区的水系设计对中国海绵城市的设计有重要的启发和指导意义（图4-23）。凡尔赛宫的主体建筑面对着山谷，山谷中心开凿了十字形水域，十字形水域周

such as wind, rain, birds, etc.[3]

In the third year after the dust has been repaved, the roof is lush, with more than 300 plant species settling here. Botanists also found some foreign species from China, India and Africa. They explained that, during the Industrial Revolution, here was a famous French textile industry base with raw materials of cotton and flax from all over the world, and these seeds were carried here in the raw materials. Some seeds have adpated to soil and climate here, so they live and reproduce.

Botanists use labels to explain species in the roof garden. Here it not only brings people historical memories beyond three-dimensional space, but also becomes a botanical garden and educational base for children to know plants. This work changes people's way of commemorating history and evaluating life. Through it, we know that exhibits in museums are of memorial significance, a classical sculpture can reappear and remodel historical culture, and landscape architecture also has its unique way to explain historical culture and modern life. People's life is changing, as well as people's understanding of society and value orientation. An excellent work may be invisible or even cannot be drawn. It needs to be jointly finished with the participation of time and nature.

Through the design of sponge city, we need to improve the ecological level of nature as well as that of social humanities, so as to provide China

③ An Jianguo and Fang Xiaoling, *Thought and Education of Landscape Architecture In France*, Beijing: Higher Education Press, 2012, P115.

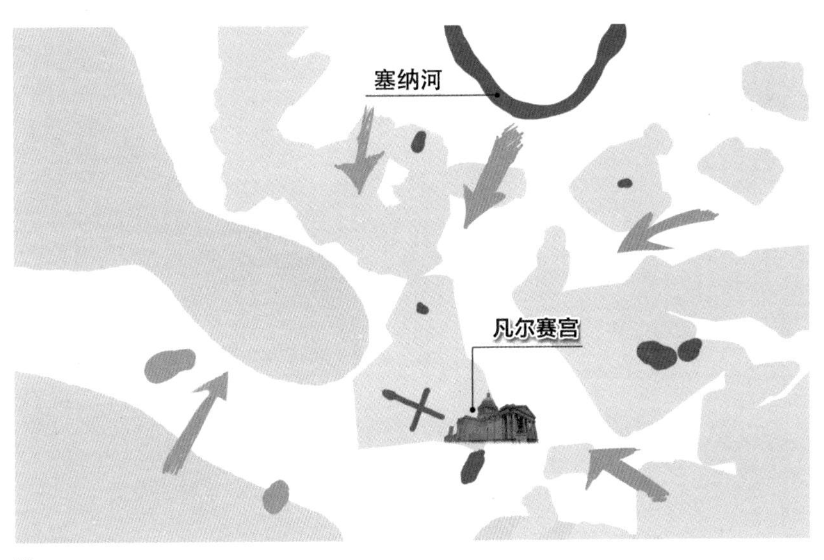

图 4-23

边是丘陵。十字形水域的水源就是这些丘陵渗流下来的雨水，只有在干枯季节才从塞纳河调水到凡尔赛宫。在雨水渗流过程中，多个特意设计的调蓄池起到了补给十字形水域水源的作用，同时也起到了调节汇水压力和滞留雨水的作用。

图4-24是为使凡尔赛宫周边丘陵汇流雨水到凡尔赛的十字形水域前面所设计的调蓄池。这些调蓄池形成自然湿地的风貌，自然生长的芦苇等水生植物对雨水进行净化。

图4-25是2012年建造的观景台，让人们从更广阔的视角宏观地看到凡尔赛宫水系的汇水方向和雨水滞留方式。如今已经呈现出湿地的调蓄池环境景色优美，水质清澈。

with its own characteristics of material and spiritual civilization in the process of sponge city construction.

3.2 Delaying

The technical point of delaying is to design all the measures that can delay rainwater. By permeating, water can be purified, groundwater can be supplied and the time that rainwater converges to river can be prolonged, so as to avoid flood peak, delay the arrival of flood peak, reduce the flow of flood peak and decrease the threat of flood.

The main technology of delaying is to use dispersed delaying measures:

(1) Buffer zone of vegetation, concave green land permeating pool, wetland;

(2) Interception, collection and use of rainwater;

(3) Ecological delaying zone;

(4) Rain garden and landscape water.

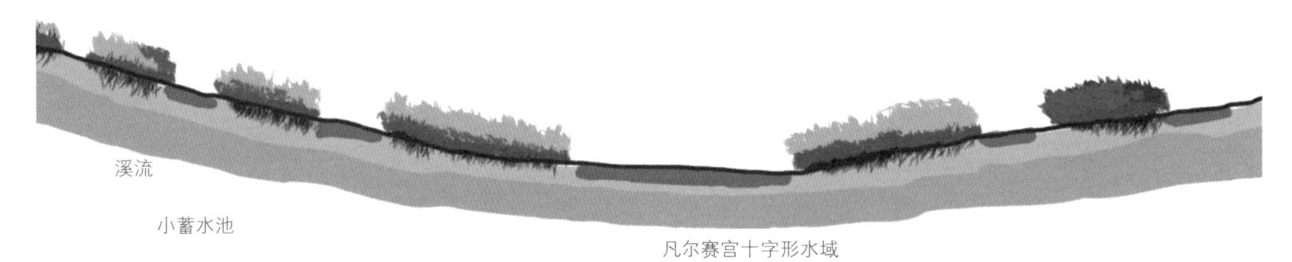

图 4-24

图 4-25

图4-26是自然的湿地水系流经凡尔赛宫周边之前，设计师设计的人工调蓄池水域。该人工调蓄池不仅有调节雨水容量的作用，也为生物营造栖息地。这些针对新的城市开发而增设的调蓄池，是凡尔赛宫新增的雨洪管理设施。

图4-27显示了山岭汇流的水系穿越城市的场景，设计师设计了容量极大的排水渠，这些排水渠的墙面都是石笼建造的。石笼下方铺设了植物土，并种植了藤蔓植物。每年洪水都会带来大量泥沙并淤积在石笼墙脚下，为植物提供丰富的营养。石笼墙体是非常完美的海绵体，它可以渗出城市中多余的雨水，也可以吸附消化一部分洪水。石笼为这些藤蔓植物创造了理想的生存环境，不必

The water system design of park zone of the Palace of Versailles in France has important inspiring and guiding significance to the sponge city design in China (Fig 4-23). The main building of Palace of Versailles faces the valley, in the center of which a cross-shaped water area was dug out. The area is surrounded by hills, and its water source is rainwater seepage from these hills. Water is transferred from the Seine to Palace of Versailles only in dry seasons. There are several regulating pools in the seepage process of rainwater. They can supply water source of the cross-shaped water area, and at the same time, regulate water convergence pressure and reserve rainwater.

Fig 4-24 shows the regulating pools designed before rainwater from the surrounding hills converging to the cross-shaped lake of the Palace of Versailles. They are now natural wetlands, with reeds and other natural aquatic plants purifying rainwater.

图 4-26

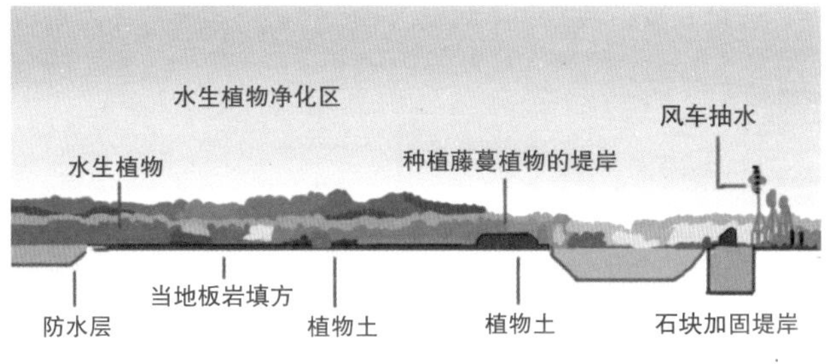

图 4–27

担心洪水会冲走植物。翌年春天，这些植物展现出生机盎然的姿态。设计师要为自然的自我调节留有空间，不必过于控制和配置植物的生长形态。

3.3 蓄

蓄的技术要点是提供一切可以让雨水储存下来的机会，以便再利用。

蓄的技术可以通过以下几种设计形式表现：

（1）地下调蓄池（蓄水池或蓄水装置）；

（2）水库、湖泊、池塘和湿地；

（3）民用建筑雨水的收集利用（雨水桶）。

3.4 净

净的技术要点是设计完善城市水循环和

Fig 4-25 is a viewing platform built in 2012, which enables people to overlook the water convergence direction and rainwater delaying ways of the Palace of Versailles water system from a higher view. The regulating pools, which appear to be wetlands today, have perfect ecological environment and clear water.

Fig 4-26 is an artificial regulating pool. Natural wetland water flows through it around the Palace of Versailles. It regulates rainwater capacity and creates biological habitat. These regulating pools that are added for new city development are a new rain-flood management measure of the Palace of Versailles.

Fig 4-27 shows the scene of water flowing through cities, which is collected from mountain range. Drains with huge capacity are designed and their walls are made of stone cages. There is plant soil at the bottom of the stone cages with vine plants planted there. Every year, flood brings a large amount of mud and sand, which silt up under the corner of stone cage wall and provide rich nutrition for plants. The stone cage wall is a perfect sponge. Extra water of the city exudes through it, and part of flood can be absorbed by it. The stone cages create ideal living environment

水环境，减少面源污染；通过植物与土壤的自净作用，生态地完成自然水净化和利用；通过人工水质处理，高标准地完成生活污水净化，改善城市水环境，改善民生，确保饮水安全。

净的技术可以通过以下几种设计形式表现：

（1）自然土壤净化；

（2）开发利用本土水生植物净化水体；

（3）城市人工污水治理。

任南琪指出：构建具有污染物净化功能的生态系统，去除部分含氮化合物和含磷化合物，科学制定污水处理厂排放标准，避免污水过度处理，实现节能降耗。防止污水直接进入城市水体，将污水全部截流，提倡排水系统采取雨污分流制，特别是在逢雨必涝的街区，雨水经过滞留和净化，就近排入城市自然水系。污水处理厂应尽可能设在内河的中上游，引入尾水以补充水资源。

如图4-28所示，法国阿赫那市生活废水处理公园用生态的方法解决了生活废水净化，实现了再生水利用。目前世界上主要有四种方法对污染进行降解：第一种是物理降解；第二种是化学降解；第三种是微生物降解；第四种是植物降解。该公园主要采用了物理降解和植物降解的方法对生活废水进行净化。

for vine plants. There is no need to worry about flood washing away plants. In next spring, these plants will be flourishing again. Designers should leave room for the self-regulation of nature, and don't need to control and arrange the growing form of plants excessively.

3.3 Storing

The technical point of "storing" is to design all chances that can store rainwater for its reuse.

The technology of "storing" can be reflected in the following design forms:

(1) Underground regulating pool (reservoir or water storage devices);

(2) Reservoir, lake, pond and wetland;

(3) Rainwater collection and use of civil construction (rainwater bucket).

3.4 Purifying

The technical point of "purifying" is to design and improve urban water cycle and water environment and reduce diffused pollution; complete the purification and utilization of natural water ecologically through the self-purification of plant and soil; through artificial water treatment, complete domestic sewage purification in a high standard, improve urban water environment and people's life, and ensure healthy water.

The technology of "purifying" can be reflected in the following design forms:

(1) Natural soil purification;

(2) Develop and use local aquatic plants to purify water;

(3) Artificial sewage system treatment of the city.

图 4-28

阿赫那市的生活废水经杂物过滤网过滤后，流经一个人工山丘，该山丘由大小不等的沙石构成，用来再过滤生活废水中的大块垃圾，然后生活废水流经栽植在细沙和黏土上的柳树林，树林在约1 m深的沙土中生长出非常密集的根系，加之细沙的渗滤作用使生活废水得到初步净化。在这一环节里，不仅生活废水得到了净化，柳树也获得了充分的营养。

生活废水经过密集的柳林后进入“之”字形水生植物区（以芦苇为主），这里主要通过植物的根系瘤和根系圈吸附生活废水中的氯化钠和重金属污染物。在植物中，芦苇的根系瘤和根系圈对污染物有很强的吸附作

Ren Nanqi pointed out that we should construct an ecosystem with pollutant purification function, which removes part of nitrogen compounds and phosphorous compounds; scientifically formulate drainage standard of sewage treatment plant, so as to avoid over-treating of sewage and realize energy saving; avoid sewage directly entering urban water and intercept all the sewage; encourage rain-sewage diversion system in drainage system, especially in the blocks where there is flood every time it rains; after being delayed and purified, rainwater drains into urban natural water system nearby; build sewage treatment plants in the midstream or upstream of inland river as much as possible, and introduce tail water as a supplement of water resources.

Fig 4-28, Harnes of France solved the problem of domestic waste water purification in ecological way, and achieved the use of recycled water. Currently,

用且生长迅速，通过光合作用可以将污染物由根系吸附到茎部转化成木质素，从而达到降解水体污染物的目的。每年入冬前，被割除的芦苇秆经过焚烧处理后重新提炼重金属，生产天然肥料和沼气，这样做也是为了避免芦苇秆腐烂于水中，重新造成污染。

但是，在植物净化区经常会出现如下问题：第一，水域表面出现很多浮萍；第二，藻类植物繁殖迅速；第三，水体缺氧；第四，水中细菌繁殖过快。

设计师设计了四个风车来解决这些问题。通过风能，风车将植物净化区的水抽向高处，再洒落在风车下面的水泥台阶上，水流经台阶最终汇入植物净化区。水由高处向低处洒落和流动，完成了水的加氧过程。水流经水泥台阶的过程，是对水的杀菌过程。因为紫外线对水的穿透力只有10 cm，加上浮萍的遮挡，植物净化区水体中的细菌繁殖很快，容易滋生耗氧量极大的藻类植物。水流经水泥台阶时的水体厚度约为2 cm，紫外线会有效地对水体进行杀菌。

处理后的再生水汇流到天然游泳池中，游泳池的水也在不断地流动，通过一片芦苇净化区，自然净化的水最终汇流到自然河道。[①]

① 安建国，方晓灵：《法国景观设计思想与教育》，北京：高等教育出版社，2012年版，第103页。

there are 4 ways of degrading pollution: physical degradation, chemical degradation, microorganism degradation and plant degradation. Physical degradation and plant degradation are mainly used in this park to purify domestic waste water.

Domestic waste water of Harnes flows through an artificial hill after being filtered by debris filter. This hill is made of gravels of different sizes, and is used to refilter large pieces of garbage in domestic waste water. Then domestic waste water flows through willow forest which is planted in sand and clay. Very dense root systems of the forest are grown in about 1-meter-deep sand. In addition, the filtering function of sand purifies domestic waste water preliminary. In this part, domestic waste water is purified and willow forest gains rich nutrition.

After passing the dense willow forest, domestic waste water enters a “zigzag” aquatic plant (mainly reeds) area. Here, root tumors and circles of plants mainly absorb sodium chloride and heavy metal contaminants in domestic waste water. Among all plants, root tumors and circles of reeds have very strong ability of absorbing contaminants. What’s more, reeds grow fast. They can absorb contaminants from roots to stems and transform them to lignin through photosynthesis, so as to achieve the goal of degrading water pollution. Every year before winter, reeds are cut and sent to incineration plant, to refine heavy metals through incineration and produce natural fertilizer and biogas. In this way, reeds cannot rot in water and cause pollution again.

However, the frequent problems in the plant purification area are as follows: First, much duckweed floats on water. Second, algae reproduces

3.5 用

用的技术要点是将收集的雨水再利用。根据不同需求和功能，所收集的雨水可以直接使用，也可以净化后使用。

用的技术可以通过以下几种设计形式表现：

（1）景观水景用水，如图 4-29a 所示；

（2）冲厕用水，如图 4-29b 所示；

（3）汽车清洗用水，如图 4-29c 所示；

（4）混凝土搅拌用水，如图 4-29d 所示；

（5）绿化灌溉和城市菜园用水，如图 4-29e 所示；

（6）环卫洗街用水，如图 4-29f 所示。

3.6 排

排的技术要点是安全排放，确保干净的雨水排入受纳水体，将合格干净的水体排出城市，最大限度地还原到自然水循环当中。

排的技术可以通过以下几种设计形式表现：

（1）雨水通过土壤自然渗透净化后排放（如过滤式沉淀槽和洼沟）；

（2）雨水通过湿地水生植物净化后排放；

（3）雨水与生活废水分流处理排放（人工净化后管网排放）。

quickly. Third, there is a shortage of oxygen in water. Fourth, bacteria in water reproduce too quickly.

Designers designed 4 windmills to solve these problems. Windmills pump water in plant purification area high through wind energy. Water spills and flows through cement steps under windmills, and finally converges to plant purification area. The oxygenation process of water is finished when water spills from high to low and flows. When water flows through cement steps, it is a sterilization process. Since ultraviolet radiation can only penetrate water for 10 centimeters, and in addition to the cover of duckweed, bacteria in plant purification area reproduce quickly, and algae that has a great consumption of oxygen emerges easily. When water is flowing through cement steps, the thickness of water is about 2 centimeters, and ultraviolet radiation can effectively sterilize it.

The recycled water after being treated converges to natural swimming pool. Water in natural swimming pool also flows constantly. Naturally purified water finally converges to natural river through a reed purification area.[①]

3.5 Using

The technical point of "using" is to reuse the collected rainwater. According to different needs and functions, the collected rainwater can be used directly or used after being purified.

The technology of "using" can be reflected in several design forms:

---

① An Jianguo and Fang Xiaoling, *Thought and Education of Landscape Architecture in France*, Beijing: Higher Education Press, 2012, P103.

图4-29a

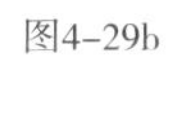

图4-29b

图4-29c

图4-29d

图4-29e

图4-29f

(1) Water used for landscape water features (Fig 4-29a);
(2) Water used for flushing toilets (Fig 4-29b);
(3) Car cleaning (Fig 4-29c);
(4) Concrete mixing (Fig 4-29d);
(5) Green irrigation/ City vegetable garden (Fig 4-29e);
(6) Cleaners washing streets (Fig 4-29f).

3.6 Draining

The technical point of "draining" is to drain safely and make sure that the water drained is clear. Qualified and clean water is drained out of city, and reverted to natural water circulation as much as possible.
The technology of "draining" can be reflected in several design forms:
(1) Drain rainwater after permeation and purification of natural soil(such as filtering precipitation tank and ditch);
(2) Drain rainwater after wetland aquatic plant purification;
(3) Drain rainwater after being separated from domestic waste water(drain through pipelines after artificial purification).

五 海绵城市建设过程中呈现出的两个水循环尺度

Two Water Cycle Scales in the Process of Sponge City Construction

1. 地域水循环尺度（包括城市在内）

在这个尺度中，城市作为大尺度地域水循环的一个环节，参与到水循环和水交换之中。被切断的水系需要重新连接起来，构建一个完整的地域水系生态系统，大尺度地域水循环直接影响海绵城市的治理效果（图5–1）。

1 Regional Water Cycle Scale (Including City)

In this scale, city as a part of large-scale regional water cycle, takes part in water cycle and exchange. The water system that has been cut off needs to be connected again to construct a complete regional water ecosystem. The large-scale regional water cycle directly influences the renovation effect of sponge city (Fig 5-1).

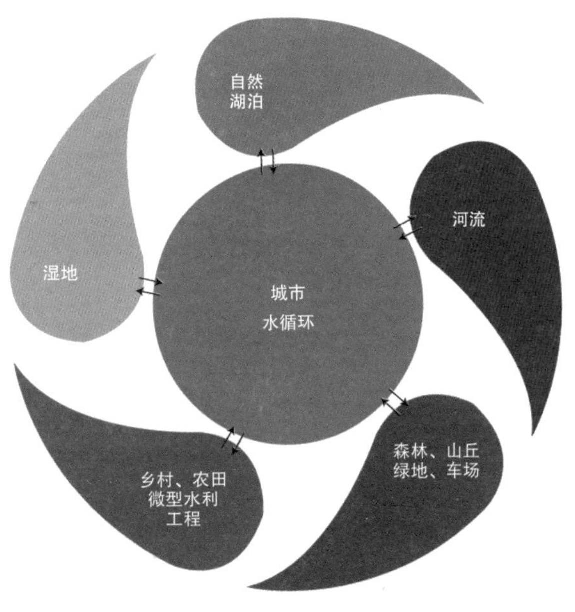

图 5–1

2. 城市水循环尺度（雨水＋污臭水系）

在这个尺度中，城市水系贯通和污臭水系治理是关键：实现地表水资源、污水资源、生态用水、自然降水和地下水等统筹管理保护和利用；充分考虑水资源、水环境、水生态、水安全和水文化；缓解城市热岛效应，确保社会水循环能够与自然水循环相互贯通（图5–2）。

2 Urban Water Cycle Scale(Rainwater and Sewage System)

In this scale, the key points are the connection of urban water system and the renovation of sewage system. We should implement the overall management, protection and use of surface water resources, sewage resources, ecological use of water, natural precipitation and groundwater, etc.; fully consider water resources, environment, ecology, security and culture; ease the heat island effect in the city and ensure the connection of social water cycle and natural water cycle (Fig 5-2).

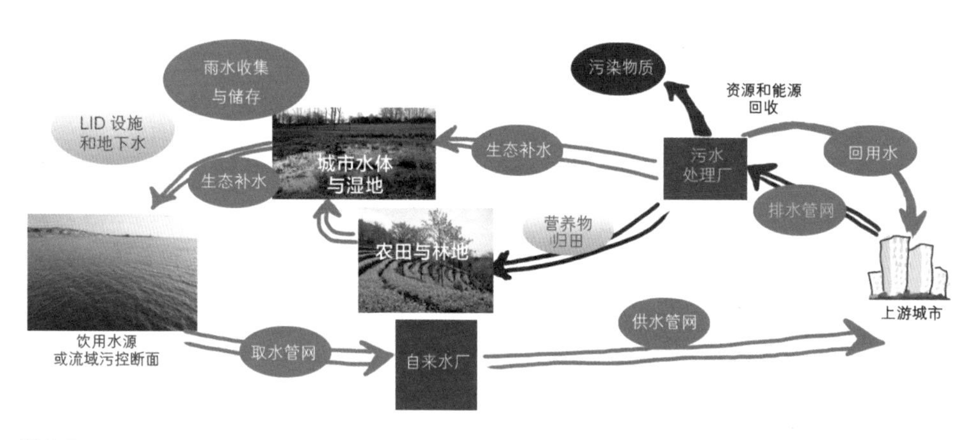

图 5–2

海绵城市实践过程中需要注意的三个问题

Three Problems That We Need to Pay Attention to in the Practice Process of Sponge City

1. 土壤污染问题

设计雨水下渗或滞留之前，要对下渗的土壤进行检测和评估，避免因雨水下渗导致土壤污染物扩散（图6–1）。

德国西部的工业区曾经是浓烟滚滚、粉尘飞扬、不见天日的钢铁基地，残留很多重金属污染物，且面积广阔，因此很难净

1 Soil Pollution

Before the design of rainwater permeating or delaying, there should be evaluation tests on the permeating soil so as to avoid the diffusion of pollutants in soil due to rainwater permeating (Fig 6-1).

The industrial zone in West Germany was once a dark steel base with a large amount of smoke and dust. Lots of heavy metal pollution was left here. The

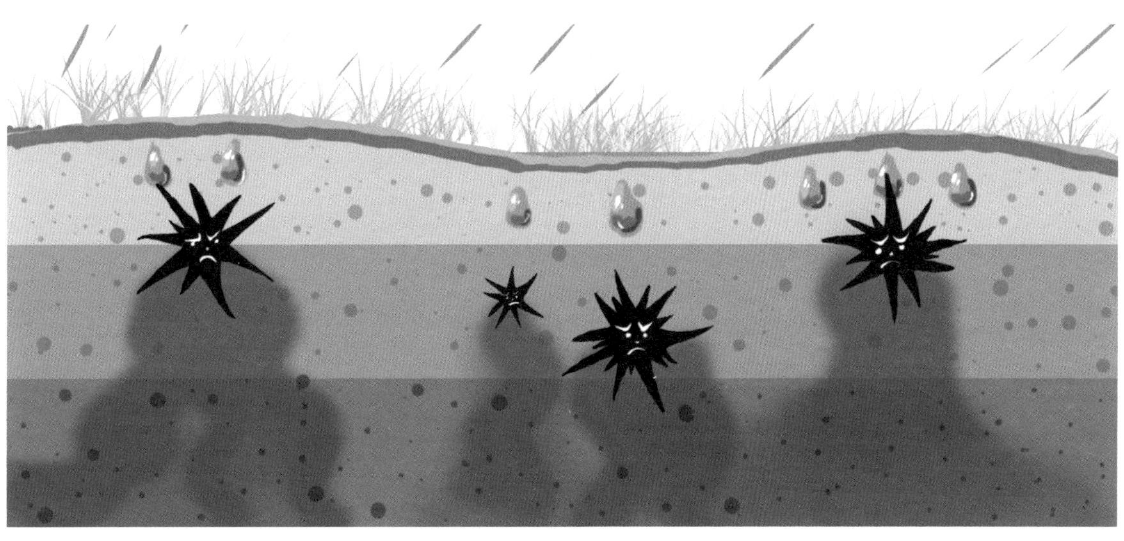

图 6–1

化。该区主要采用两种方法处理土壤污染问题：

第一种方法是将污染土壤挖出来堆在高处，用防水膜覆盖后，再加铺30 cm~50 cm的植物土，以便培植草坪。这样，污染土被罩住并固定在高处，雨水或地下水不会接触到污染土，从而达到封锁污染物，防止其扩散的目的（图6–2）。

第二种方法是对特大面积的污染土地采用“地下隔离墙”，来阻止污染的土壤与地下水接触。具体操作方法是，在污染的土壤周边挖15 m深的壕沟并灌注水泥，将污染土围起来，然后在地表铺设防水层，避免雨水径流渗入土壤污染区扩散污染。浇筑15 m深的地下水泥墙，是为了阻隔安息

polluted area was so big that it was hard to clear. Two methods were mainly used in this zone to solve soil pollution:

The first was to dig out the polluted soil and pile it high, cover it with waterproof membrane and pave 30 cm~50 cm of plant soil so as to cultivate lawn. In this way, polluted soil was covered and fixed in a high place, without being touched by rainwater or groundwater, so that the pollutants were locked and would not disperse (Fig 6-2).

The second was to use the method of “underground isolation wall” for the extra large area of polluted land, to cut off the contact between polluted soil and groundwater. Specific operations were: dig a 15-meter-deep ditch around the polluted soil and pour cement into it to encircle the polluted land. Pave waterproof layer on land surface to avoid rainwater runoff permeating into polluted area and dispersing pollution. Pouring a 15-meter-deep

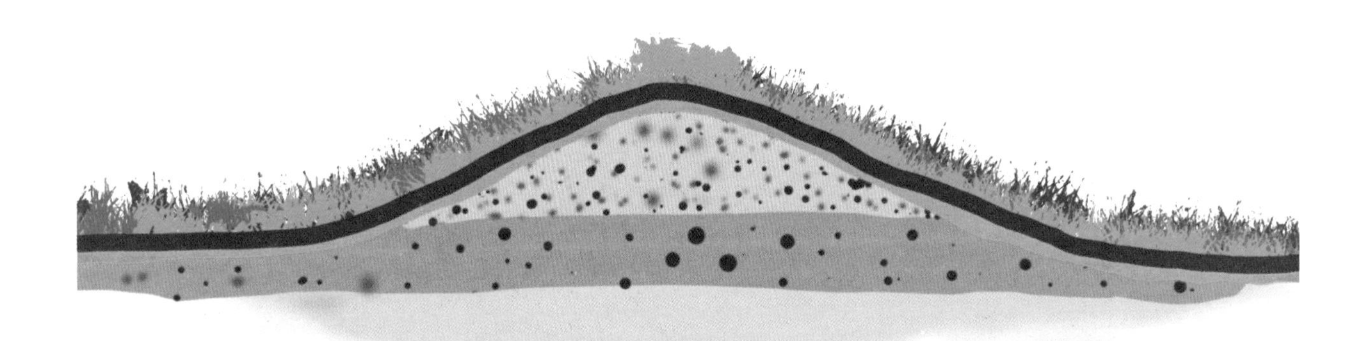

图 6–2

香扩散，钢铁工业区中的安息香污染源可达12 m深，而该物质也是最致命的污染物之一。根据其密度，安息香将悬浮在地下12 m深左右，用这种锁定法可以避免安息香在地下水中扩散（图6–3）。[①]

underground cement wall was to cut off the diffusion of benzoin since the pollution density of benzoin in the steel industrial area could be as high as 12 meters and it was one of the most deadly pollutants. Benzoin would suspend at about 12-meter-deep underground according to its density, so this method could avoid its diffusion in groundwater (Fig 6-3).[①]

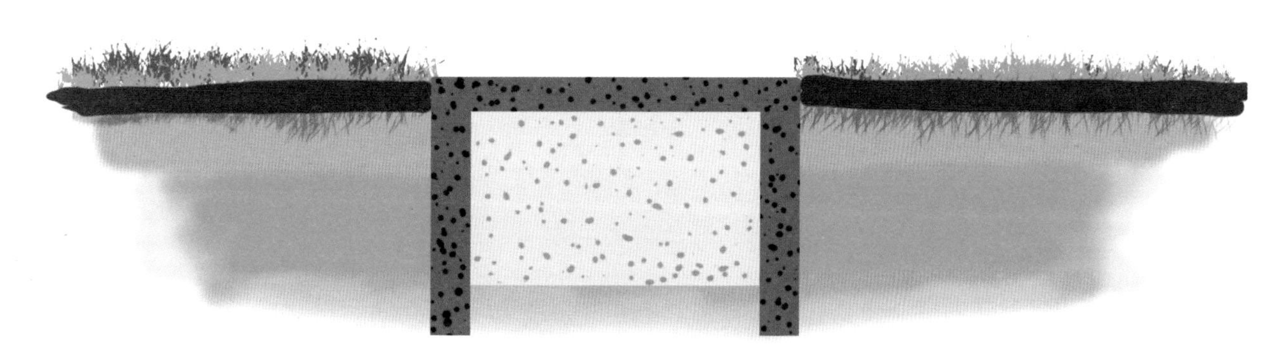

图 6–3

2. 地下水源污染问题

雨水下渗设计之前，要对地下水源进行检测和评估。如果地下水源已经受到污染，则要先考虑治理地下水污染，然后再进行海绵体设计，避免因雨水下渗扩散到地下造成的水源污染（图6–4）。

2 Groundwater Source Pollution

Before the design of rainwater permeating, there should be evaluation tests on groundwater source. If it has been polluted, designers should consider eliminating groundwater pollution before designing sponge, so as to avoid the diffusion of pollution in groundwater due to rainwater permeating (Fig 6-4).

① 安建国，方晓灵：《法国景观设计思想与教育》，北京：高等教育出版社，2012年版，第83页。

① An Jianguo and Fang Xiaoling, *Thought and Education of Landscape Architecture in France*, Beijing: Higher Education Press, 2012, P83.

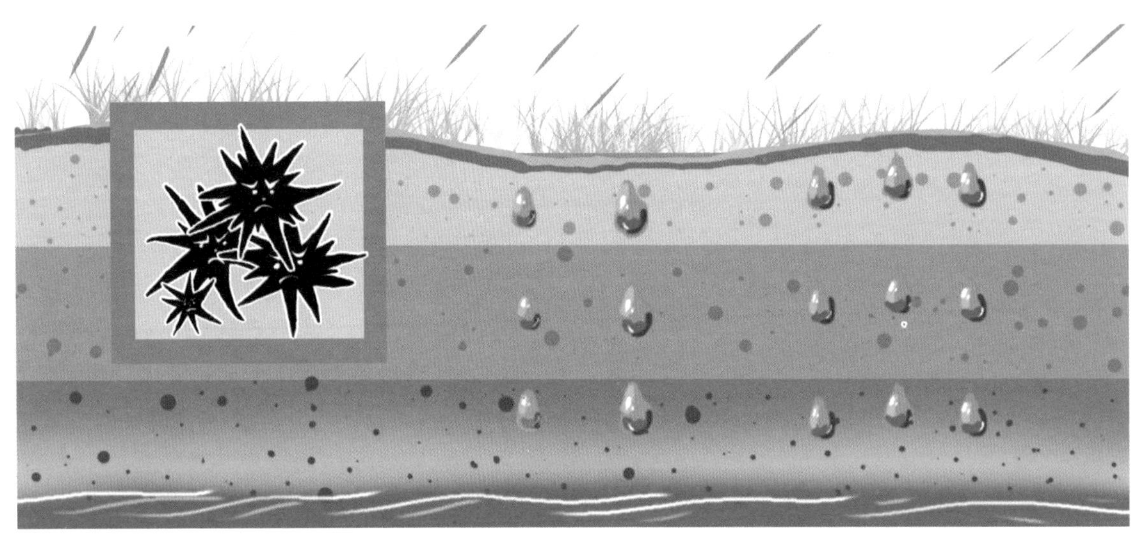

图 6-4

3. 城市黑臭水体的治理问题

任南琪指出：许多城市水体受到了污染，水体黑臭。个别城市污水全部截流后无生态补水，不得不引入清洁的水资源，但这一措施不可持续，城市水体自然生态净化功能没有得到充分利用。

治理城市黑臭水体有如下几个重要条件（图6-5）：

（1）具有足够的水动力。补水充足，流动性好，自然或人工强化复氧能力强；

（2）持续的生态净化系统与措施。建立强化自然的人工处理系统，形成自然水生生态环境与景观；

3 Renovation of Urban Black Water and Smelly Water

Ren Nanqi pointed out that the water body of many cities was polluted, which was black and smelly. In some cities, there is no ecological water supplement after the sewage is totally stopped. Therefore, clean water has to be introduced, which is unsustainable. The natural ecological purification function of urban water has not been fully used.

Important conditions for the renovation of urban black and smelly water are as follows (Fig 6-5):

(1) There should be enough water power, which means sufficient water supply, good fluidity and strong natural or artificial reoxygenation capacity of water;

(2) Sustainable ecological purification system and measures, which mean constructing artificial treatment system that is naturally strengthened, and

（3）初期修复的技术措施。清除底泥与修复，快速处理有机和无机污染物；

（4）严格控制点、面污染源。污水全部截流，达到一定排放标准，净化与调控点、面污染源。

forming natural water ecological environment and landscape;

(3) Technical measures of renovation in initial stage, which means clearing and repairing sedimentation, quickly treating organic and inorganic pollutants;

(4) Strictly control of point and surface pollution sources, which means stopping all sewage and achieving a certain drainage standard, and purifying and regulating point and surface pollution source.

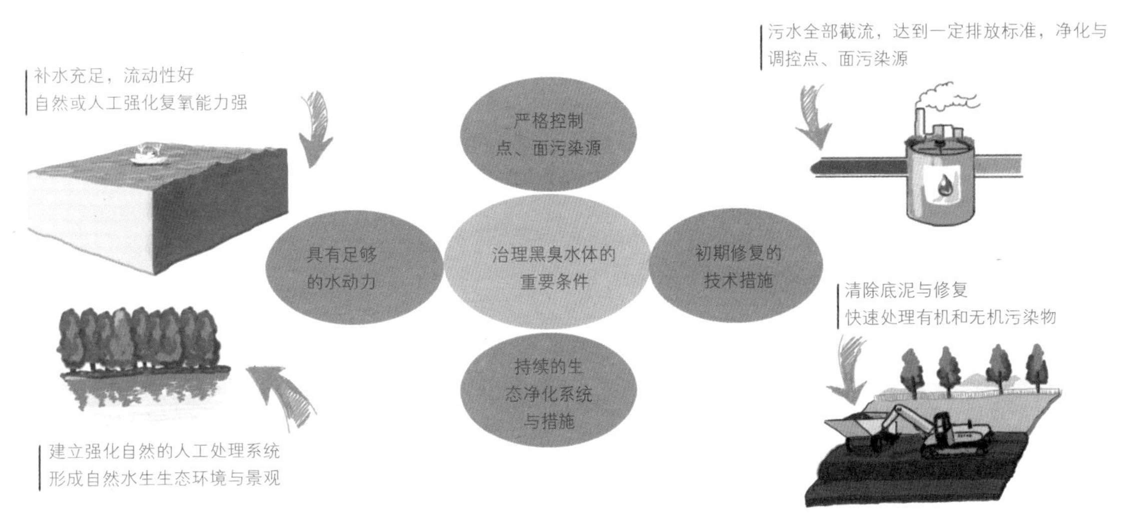

图 6–5

七

中国的海绵城市与其他国家相关理论的区别

The Differences Between Theories of Sponge City in China and in Other Countries

1. GSI / GI（绿色基础设施）

绿色基础设施指一个相互联系的绿色空间网络，由各种开敞空间和自然区域组成，包括绿道、湿地、雨水花园、森林和乡土植被，这些要素组成一个相互联系、有机统一的网络系统。该系统可为野生动物迁徙和生态过程提供起点和终点，还可以管理暴雨，减少洪水危害，改善水的质量，节约城市管理成本。①

2. LID（低影响开发）

低影响开发指采取土地规划和工程设计的方法来管理雨水径流。它强调现场自然属性的保持和利用，从而保护水质。②

1 GSI/GI (Green Stormwater Infrastructure/ Green Infrastructure)

Green stormwater infrastructure, refers to an interconnected green space network which is composed of various wide open spaces and natural areas including elements such as green roads, wetlands, rain gardens, forests and native vegetation that comprise an interconnected, organic and unified network system. The system provides a starting point and an end point for wildlife migration and ecological process and can manage the rainstorm, reduce the danger of flooding, improve the quality of water and save urban management costs. ①

2 LID (Low-Impact Development)

LID is to manage rainwater runoff by means of land planning and engineering design. LID emphasizes the maintenance and utilization of natural attributes on the spot, so as to protect water quality. ②

①② 安建国，方晓灵：《法国景观设计思想与教育》，北京：高等教育出版社，2012年版，第83页。

①② An Jianguo and Fang Xiaoling, *Thought and Education of Landscape Architecture in France*, Beijing: Higher Education Press, 2012, P83.

3. WSUD（水敏性城市设计）

水敏性城市设计指城市设计与城市水循环的管理、保护和保存的结合，从而确保城市水循环管理能够尊重自然水循环和生态过程。[3]

4. SUDS（可持续排水系统）

建立可持续排水系统的目的是减少新发展和现有发展对地表排水的潜在影响。“可持续城市排水系统”这一术语还未被认可，为适应农村水管理的可持续性实践，“城市”这一词已被删除。

起初，可持续城市排水系统是英国对可持续城市排水系统所采取的举措。但这个举措不仅仅应用在“城市”里，因此，为了减少误解，通常将“城市”一词删除。其他国家在用词上也采用了类似的方法，比如，美国最佳管理实践（BMP）与低影响开发和澳大利亚水敏性城市设计。[4]

图7-1展示了水治理的历史发展过程。

中国海绵城市理论的提出属于以上四种理论的延续，但是中国独特的国情条件，使中国面临着更加艰巨的城市环境改造任务：

③ 安建国，方晓灵：《法国景观设计思想与教育》，北京：高等教育出版社，2012年版，第83页。

④ 引自维基百科

3 WSUD (Water Sensitive Urban Design)

Water Sensitive Urban Design is the combination of urban design and management, protection and preservation of urban water cycle, thus ensuring that urban water cycle management can respect natural water cycle and ecological process.[3]

4 SUDS (Sustainable Drainage System)

A sustainable drainage system (SUDS) is designed to reduce the potential impact of new and existing developments on surface water drainage discharges. The term “Sustainable Urban Drainage System” has not been accepted yet. “Urban” has been removed so as to accommodate rural sustainable water management practices.

Originally the term SUDS was used to describe the UK approach to sustainable urban drainage systems. These measures may not necessarily be in “urban” areas, and thus the “Urban” part of SUDS is now usually deleted to avoid confusion. Other countries adopt similar approaches in terms of terminology, such as Best Management Practice (BMP) and Low-Impact Development in the United States and Water-Sensitive Urban Design in Australia.[4]

Fig 7-1 shows the historical dovelopment process of water management.

The theory of sponge city in China is put forward as the continuance of the above 4 theories. However, China is facing the harder task of urban environment

③ An Jianguo and Fang Xiaoling, *Thought and Education of Landscape Architecture in France*, Beijing: Higher Education Press, 2012, P83.

④ From Wikipedia

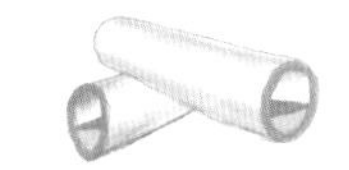

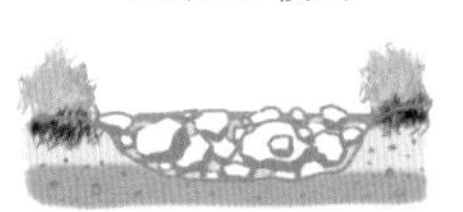

图 7–1

针对土地和物候的特殊性对水资源、水环境、水生态、水安全、水文化进行系统的设计。

中国海绵城市改造的特殊背景：

· 城市尺度普遍大于国外

· 城市污染较国外一些城市严重（生活垃圾、生活废水、工业废水、土壤污染和地下水污染等）

· 中国城市缺乏生物多样性，生态系统修复更困难

· 城市绿化与城市居住人口的比例严重失衡，城市人均绿化面积过小，人口数量与植物数量比例不匹配，应尽可能地增大绿化面积

renovation because of her unique national conditions: to make a systematic design of water resource, environment, ecology, security and culture according to the particularity of land and phenology.

Special background of sponge city renovation in China:

· Urban scale of China is usually larger than that of foreign countries

· Pollution in Chinese cities is more serious than that in foreign cities (domestic waste, domestic waste water, industrial waste water, soil pollution, groundwater pollution, etc.)

· The lack of biological diversity in Chinese cities, which makes ecosystem restoration even harder

· The proportion of city greening and city residents is seriously unbalanced. Per capita green area in the city is too small, and the tree-planting proportion does not match the population. Green area should be increased as much as possible

八

海绵城市设计前的数据采集

Data Collection Before the Design of Sponge City

海绵城市的设计需要诸多科学数据的支撑，综合学科的交叉运用是海绵城市设计的特征。这项工作的推进，使规划设计真正具备了理性与感性的双重特征。

主要的理性数据：

·地理地质特征介绍

·针对海绵城市敏感地块的评估和地域土壤分类研究：

(a) 透水率高的海绵土层（地下渗水蓄水能力数据预估）

(b) 透水率低的土层或岩石层，可做调蓄池蓄水再利用

·植物蓄水能力数据预估（各种植物的蓄水能力参数）

·植被蒸发量数据

·降雨量区域测评历史数据等水文信息

·主要水系汇流方式和汇水流量数据

·主要水系经过主要城市的汇流和时间数据

The design of sponge city needs the support of many scientific data, and the cross use of comprehensive subjects is its feature. The advancement of this work makes planning and design actually have dual characteristics of rationality and sensibility.

Main rational data:

·An Introduction to Geographical and Geological Characteristics

·An Evaluation on Sensitive Plots of Sponge City/ Study on Classification of Regional Soil

(a) Sponge soil layer with high water permea-bility (Data prediction of underground water permeation and storage capability)

(b) Soil layer or rock layer with low water permea-bility, which can be reused as regulating pool

·Data Evaluation on Water Storage Capacity of Plants (Parameter of water storage capacity of each plant species)

·Data of Vegetation Evaporation

·Hydrological Information such as the Historical Data of Rainfall Area Evaluation

·Data of Convergence Way and Flow of Main Water Systems

·Data of Convergence and Time of Main Water Systems Flowing Through Main Cities

· 现有主要城市地下管网排水量设定数据

· 地表蒸发量数据（不同材料在雨季常规温度下的蒸发量计算）

· 城市发展数据（交通系统构建过程、城市形态和建筑群落构建过程、绿地演化过程、人口变化分析、工业分布和影响、各类污染数据和原因等）

理性数据是海绵城市设计的重要依据，但在理性数据的基础上，必须有感性数据的校正。感性对空间的理解有其独特性，是对理性空间理解的有益补充。

我们总是能够通过感知来发现那些与现实相关的特殊关系，这些特殊关系指被我们的常规观察忽略掉，但又与现实密切相关的信息。比如，在海滩漫步时的愉悦，可能更敏感的是脚下皮肤与海水、细沙接触产生的愉悦，这种触觉带来的空间理解与现实关系密切。我们每次在现场的感受都不尽相同，把这种由时间、空间和现场心情等导致的不同感性数据与理性数据结合起来思考，会成为设计的重要灵感来源。

通过感知，身体的经验与我们生活的世界可以进行对话。我们可以不断地用身体去验证更大的空间尺度，而非让大尺度的土地空间模糊了我们对场地的真实理解。这表明我们可以超越身体尺度去理解地域尺度。

· Data of Underground Pipelines Drainage Amount of Current Main Cities

· Data of Land Surface Evaporation(calculating the evaporation of different materials at normal temperature in rainy season)

· Data of City Development (construction process of transportation system, construction process of urban form and architectural community, evolution process of green land, analysis of population change, industrial distribution and impact, various pollution data and reasons, etc.)

Rational data are important reference to sponge city design, but there must be correction of emotional data on the basis of rational data. The understanding of emotional space has its own uniqueness, and it is a useful complement of rational space.

We can always notice the special relationship related to reality through perception. These special relationships often refer to information that is ignored by routine observation, but is closely related to reality, such as the pleasure of wandering on the beach. The pleasure of feet touching the beach, seawater and sand may be more sensitive. The space understanding brought by such sense of touch has a close relationship with reality. What we feel every time on spot is not the same. Combining such kind of different perceptual data due to time, space and mood with rational data in thinking will become an important source of inspiration of design.

By perception, body experience can interact with the world in which we live. We can continually measure larger spatial scales through our body, rather than let the large-scale land spaces obscure our real understanding of the site. It means that we can understand the regional scale beyond the body scale.

In addition, we cannot understand the landscape

另外，我们是无法只通过鸟瞰图和地图去理解环境的，必须脚踏实地不断移动去理解空间，我们有必要重新发现并标注图纸、地图疏漏的感知信息。景观设计师的工作是通过不同尺度的感知研究（感性部分），串联起其他景观所需的科学领域研究（理性部分）。例如，通过地理信息系统的数据，我们可以得出山体公园中最佳的步行路线。但是，只有感性体验才能发现这些理性数据的局限性。这种局限性表现在大数据测算与实际空间感受有出入，我们虽然可以根据地理信息系统数据选择最佳山体游览路线，但有些重要的景点在计算的路线中不可见，或者观看位置不佳，这就必须通过人体的感知系统来校正这些理性数据的缺陷。设计师会通过现场观测来定位最佳的观景点，根据视域需求来修剪和种植植物；会通过嗅觉的愉悦来设计停留的空间和逗留的时间；会通过触觉体验在空间中获得的场地信息；会通过听觉去理解非视觉空间表达的空间和心理感应；有时还会通过味觉去记忆特殊的场所。即使是“凝思”和“发呆”，也是场地给予了某种客观条件才可以促成的行为。所以，这种以人的尺度为校正标准的感知方法在规划设计中越来越重要。这种肢体感知的方法可以在任何规划尺度下应用，因为人的移动会改变和增加对空间洞察的信息。

solely through bird’s eye view and maps. Instead, we must come down to earth and try to understand the space through constant movement. It is necessary that we point out and relabel the missing perception information on plans and maps. The work of landscape architects is to connect other studies of scientific fields (rational) required by landscapes through perception studies (emotional) in various scales. For example, we can find the best walking route in the mountain park through the data of geographic information system. However, only by emotional experience can we find the limitation of the rational data. Such limitation appears as the discrepancy of big data calculation and actual space feeling. Although we can choose the best mountain tour route according to the data of geographic information system, some important scenic spots are not in the calculated route, or the viewpoint is not good. Therefore, the defects of these rational data must be corrected by human perception system. Designers will locate the best viewpoint through sense of vision, and trim and plant plants according to the requirements of vision field; they will design the space and time of staying through the pleasure of sense of smell; they will experience the spot information that users will gain in the space through the sense of touch; they will understand spacial and psychological induction that cannot be expressed in visual space through sense of hearing; sometimes they will remember special places through sense of taste. Even “meditation” and “daze” are both triggered by the certain objective factor of the spot. Therefore, such a perception method with human scale as the correction standard is more and more important in planning and design. The method of limb perception can be applied in any planning scale

作者提倡应用“感知引领设计”的方法对设计的场地空间和使用者的使用方式进行感性的信息搜索，将这些看似模糊的信息通过理性数据的论证，使其形成独特的、因地制宜的规划和设计思路。这种方法优于“模式化设计”的方法，因为它帮助设计师拓宽思路，找到设计的真谛。

because the movement of humans will change and enlarge the information of the space.

The author of the book promotes the method of “perception leading design”, which does emotional information search for site space of the design and using manner of users, then makes these seemingly obscure information into unique planning and design ideas according to local conditions through the demonstration of rational data. This method is better than “modular design”. It inspires designers to find the real meaning of design.

九 海绵城市施工后的数据追踪

Data Tracking After the Construction of Sponge City

目前，海绵城市的建设工作还处于探索阶段，各个地区海绵城市的案例实施需要一个完整的数据追踪过程，以便建立数据库，为以后的海绵城市数据化和量化标准的研究提供科学支持。这种数据追踪主要包括海绵体蓄水数据、渗水数据和水质数据的追踪统计。

· 海绵体修复后的蓄水数据统计（雨水花园、城市滞留池、植草沟等）

· 海绵体修复后的渗水数据统计（原有土壤、改良后的土壤、地面铺装材料等）

· 海绵体修复后的水质数据统计（自然降雨、城市水系）

通过这些具体的海绵体数据追踪以及水文数据对比，我们可以获得第一手海绵城市

At present, the construction of sponge city is still in exploration. Implementation of sponge city cases in different areas needs a complete data tracking process, so as to build a database and provide scientific support for the research of sponge city digitization and quantification standards in the future. Such data tracking is mainly reflected in the tracking and statistics of water storage data, water permeation data and water quality data of sponge.

· Water Storage Data Statistics After Sponge Renovation (rain garden, city retention pool and grass ditch, etc.)

· Water Permeation Data Statistics After Sponge Renovation (original soil, improved soil and paving materials, etc.)

· Water Quality Data Statistics After Sponge Renovation (natural rainfall and urban water system)

Through these specific sponge data tracking and hydrological data comparison, we can get firsthand

实施效果的参数。通过对这些参数进行长期比较，海绵城市的设计会越来越科学、越来越可控。

parameter of sponge city implementation effect. After years of comparison, these data will finally make sponge city design more and more scientific and controllable.

因地制宜地设计海绵城市

Design Sponge City According to Local Conditions

1. 我国海绵城市建设的四个关注

1.1 关注不同城市的特殊性

因物候和地理环境的不同，每个地区、每个城市都有其特殊性。海绵城市的设计与实施需要适应这种差异性。如，中国东北和内蒙古有冰期，其雨水和水系管理具有明显的地域性特征；山西的失陷性黄土是该地区雨水与土壤问题的焦点。因此，因地制宜地研究并解决不同地区的海绵城市问题是必要的，应避免模式化套用，导致损失和资源浪费。

1.2 关注模式化思维对海绵城市建设的负面影响

上文提到的模式化经验，俗称“典范”。对于以土地构建为基本特征的海绵城市项目来说，模式化的作用微乎其微。每一个地区的成功经验都离不开该地区的土地和物候特征，而这些特征恰恰是不可复制的，

1 Four Points That Sponge City Design Pays Attention to in China

1.1 Pay Attention to the Particularity of Different Cities

Every area or every city has its particularity due to the differences in phenology and geographical environment. The design and implementation of sponge city need to adapt to these differences. For example, the glacial characteristics of Northeast China and Inner Mongolia make rainwater and water system management with obvious regional characteristics; the collapsible loess in Shanxi Province is also the focus of rain and soil problems in that area. Therefore, it is necessary to research and solve sponge city problems in different areas according to local conditions, and avoid stereotyped application, which may lead to loss and waste of resources.

1.2 Pay Attention to the Negative Impact of Stereotyped Thinking on Sponge City Construction

The stereotyped experience mentioned above is commonly known as a “model”. For sponge city project taking landscape architecture as basic

也是无法作为模板推广的。因此模式化建设思维可能对海绵城市的建设产生负面影响，可能导致僵化的复制，甚至破坏当地海绵体的原有特性。因此，在统一海绵城市理论和实践的框架思路后，要因地制宜地具体研究和实施。

1.3 关注研究深度，避免急功近利

海绵城市的研究是一个世纪课题，带领着中国创造更加和谐的人居环境。在这一历史性的阶段，出现的问题也具有其阶段性特征。因此，不论是对于学者还是工作在第一线的设计师来说，都需要严谨的研究和探索精神，循序渐进地推进海绵城市研究工作，对于每个出现的新问题都要以全局观（地域的、国家的、全球的）来研究和探索解决途径。

1.4 关注源头控制性措施研究

海绵城市面临着重要的雨水净化和污臭水系治理问题。在当今人口密度普遍过高的中国城市，控制源头污染比治理污染更加重要。例如：一个烟头可以污染40吨水，吸烟者经常无意识地将烟头扔进排水沟和河道中，一秒钟完成的污染行为，其净化过程却需要数天的时间和昂贵的资金。道德修养和

characteristics, the function of a model is next to nothing, because the success experience of every area is inseparable from its land and phenological features, which cannot be copied or popularized as a model. Therefore, stereotyped construction thinking may have negative impact on sponge city construction. It may lead to rigid reproduction and even damage the original characteristics of local sponge. So after integrating the theoretical and practical framework and idea of sponge city, the construction should be specifically researched and implemented according to local conditions.

1.3 Pay Attention to the Depth of Research and Avoid Quick Success

The research of sponge city is a century topic, which is creating a more harmonious living environment. In this historical phase, the problems that occurred have phased features. Thus, both scholars and designers working on the first line need rigorous research and exploration spirit, to carry forward the research of sponge city step by step, to study and explore solutions to every new problem based on the overall (regional, national and global) situation.

1.4 Pay Attention to the Research of Source Controlling Measures

Sponge city deals with important problems of rainwater purification and sewage system renovation. In current Chinese cities, population density is generally too high, so controlling pollution at source is more important than pollution renovation. For example, a cigarette end can pollute 40 tons of water. Smokers often throw them into gutters and rivers

伦理水平的提升是海绵城市建设得以实现的重要支柱。

unconsciously. A pollution behavior is completed in a second, but the purification process needs many days and a large sum of money. The improvement of moral cultivation and ethical level is an important pillar for the construction of sponge city.

2. 中国改革开放以后的曲折探索

20世纪80年代，改革开放初期，在经济发展过程中，人们对环境问题的关注不足。

20世纪90年代，城市化进程开始，此时是环境急剧恶化的时期。

21世纪初始，污染的代价开始在各领域显现。

2010年以后，人们要求喝上一口纯净的水，吸上一口新鲜的空气，住在一个宜居的生活环境中。这使景观设计在中国获得了孕育和诞生的土壤。中国新的社会变革发生后，景观设计的研究体系和实践领域会得到确定，最终将以“土地构建设计”的身份，开启真正符合中国当代国情的理论研究与实践。在各个学科不断介入的过程中，学者应本着更包容、更开放的态度，共同完成新时代的学科建设和发展。

2 Twists and Turns After Reform and Opening Up in China

In the 1980s, the initial stage of Reform and Opening Up, there was a lack of attention to environmental issues in economic development process.

In the 1990s, the environment deteriorated drastically when the urbanization process was propelled emphatically.

In the 2000s, the cost of pollution began to appear in various fields.

After the year of 2010, people ask for drinking clean water, breathing fresh air and staying in a livable living environment, thus promoting the birth of landscape architecture in China. Landscape architecture took part in the new social changes in China, and after that, its research system and practice area will be determined with the identity of “Land Construction”, thus starting a theoretical research and practice that is really in line with China’s current national conditions. In the process that various subjects are constantly involved in, scholars should jointly complete the subject construction and development in new era with a more inclusive and more open attitude.

3. 海绵城市的理论核心

海绵城市设计的关键不在城市本身，而是通过三个尺度的设计：大尺度的地域雨水径流控制、中尺度的地下管道排蓄疏导、小尺度的城市规划设计，来解决比城市更大

3 Theoretical Core of Sponge City

The key point of sponge city design is not the city itself. It is solved by the 3 scales of design (large-

的地域自然环境和水系环境问题。必须兼顾（包括城市在内的）区域性的土地构建设计和雨洪管理，才能为海绵城市的实现奠定理论和实践基础。

海绵城市的建设不可能，也不应该用同样的方法解决，应以地域土地研究和地域水系研究为基本依托，因地制宜地设计与实施。每一个地区、每一个城市都有其特殊性，只有针对其土地和物候的特殊性对当地的水资源、水环境、水生态、水安全、水文化进行系统的设计，才能准确地完成海绵城市的治理目标。

scale regional rainwater runoff control, middle-scale underground pipeline draining and dredging, small-scale city planning and design) in the regional natural environment and water system environment, which is larger than city. Meanwhile, researchers must give consideration to regional (including city) land construction design and rain-flood management as well so that they can lay the theoretical and practical foundation for the realization of sponge city.

Sponge city construction cannot and also should not be solved with the same method. It should be designed and implemented according to local conditions, based on the regional land and water system research. Every area or city has its particularity. Only making a systematical design of local water resource, environment, ecology, security and culture according to the particularity of land and phenology, can designers correctly achieve the goal of sponge city renovation.

鸣谢

主要参与编辑人员：

孙媛媛、刘　幸、韩　骁、徐家亮、彭世伟、
李　兵、王　君、高静华、汪　杰、陈　侠、
邹裕波、闫宝生、韦爽真、李润楠、龙　赟

鸣谢：

法国高广艺术、文学与科学院
法国Atelier Mediterraneen设计集团
法国An Architecte-Paysagiste景观机构
北京大学建筑与景观设计学院
上海市政工程设计研究总院景观规划设计研究院
哈尔滨市嵘森园林景观工程设计集团有限公司
北京道勤创景规划设计院有限公司
天津雅蓝景观设计工程有限公司
陕西水石合景观规划设计有限公司
阿普贝思（北京）建筑景观设计咨询有限公司
重庆纬图景观设计有限公司
重庆尚源建筑景观设计有限公司
山西鼎景建筑景观设计有限公司
广州山水比德景观设计有限公司
成都景虎景观设计有限责任公司

Acknowledgements

Chief Editorial Staff:
Sun Yuanyuan, Liu Xing, Han Xiao, Xu Jialiang,
Peng Shiwei, Li Bing, Wang Jun, Gao Jinghua, Wang Jie,
Chen Xia, Zou Yubo, Yan Baosheng, Wei Shuangzhen,
Li Runnan, Long Yun

Acknowledgements:
Academie des Hauts Cantons (Art, Sciences et Belles Lettres)
Atelier Mediterraneen de L'Environnement
An Architecte-Paysagiste du Projet
College of Architecture and Landscape Architecture of Peking University
Shanghai Municipal Engineering Design Institute-Landscape Design
Institute
Harbin Rong Sen Landscape Architecture Engineering Design Group
Co., Ltd
Beijing Daoqin Landscape Planning Institute Co., Ltd
Tianjin Ya Lan Landscape Design Engineering Co., Ltd
Shaanxi Shuishihe Landscape Planning and Design Co., Ltd
U.P.Space (Beijing) Landscape Architecture Design Consultants Co., Ltd
Chongqing WISTO Landscape Design Co., Ltd
Chongqing A&N Shangyuan International
Shanxi Ding Jing Landscape Architecture Design Co., Ltd
Guangzhou S.P.I. Landscape Design Co., Ltd
Chengdu LANDHOO Landscape Design Co., Ltd

图书在版编目(CIP)数据

海绵城市与土地构建 / (法) 安建国，李锦生，钟律著；李亚迪译. -- 大连：大连理工大学出版社，2019.10

ISBN 978-7-5685-2388-2

Ⅰ. ①海… Ⅱ. ①安… ②李… ③钟… ④李… Ⅲ. ①城市建设—研究 Ⅳ. ①TU984

中国版本图书馆CIP数据核字（2019）第235201号

出版发行：大连理工大学出版社
（地址：大连市软件园路80号 邮编：116023）
印　　刷：深圳市龙辉印刷有限公司
幅面尺寸：225mm×290mm
印　　张：6.75
字　　数：80千字
出版时间：2019年10月第1版
印刷时间：2019年10月第1次印刷
策划编辑：苗慧珠
责任编辑：徐　丹
责任校对：曹静宜
封面设计：洪震彪

ISBN 978-7-5685-2388-2
定　　价：118.00元

电　　话：0411-84708842
传　　真：0411-84701466
邮　　购：0411-84708943
E-mail：landscape@dutp.cn
URL：http://dutp.dlut.edu.cn

本书如有印装质量问题，请与我社发行部联系更换。